THIRTEENTH EDITION

GENETICS

Laboratory Investigations

THOMAS R. MERTENS
Ball State University

ROBERT L. HAMMERSMITH
Ball State University

PEARSON

Prentice
Hall

Upper Saddle River, New Jersey 07458

Library of Congress Cataloging-in-Publication Data

Mertens, Thomas Robert
    Genetics : laboratory investigations / Thomas R. Mertens, Robert L.
Hammersmith. — 13th ed.
    p. cm.
    Includes bibliographical references.
    ISBN 0-13-174252-3
    1. Genetics—Laboratory manuals. I. Hammersmith, Robert L. II. Title.
QH440.5.M47 2007
576.5078—dc22

2006044323

Executive Editor: Gary Carlson
Editor in Chief, Science: Dan Kaveney
Project Manager: Crissy Dudonis
Purchasing Manager: Alexis Heydt-Long
Buyer: Alan Fischer
Production Editor: Cindy Miller
Art Director: Jayne Conte
Cover Designer: Bruce Kenselaar
Cover Image: Daniel L. Van Dyke, Ph.D., FACMG, Mayo Clinic, Rochester, Minnesota
Managing Editor Art Management: Abigail Bass
Art Production Editor: Rhonda Aversa
Editorial Assistant: Michael Eubanks

Any uncredited images in this text belong to the authors.

© 2007, 2001, 1998, 1995, 1991, 1985, 1980, 1975, 1970, 1964, 1960, 1956, 1952 Pearson Education, Inc.
Pearson Prentice Hall
Pearson Education, Inc.
Upper Saddle River, NJ 07458

Pearson Prentice Hall™ is a trademark of Pearson Education, Inc.

Printed in the United States of America
10 9 8 7 6 5 4 3 2

ISBN 0-13-174252-3

Pearson Education LTD., *London*
Pearson Education Australia PTY, Limited, *Sydney*
Pearson Education Singapore, Pte. Ltd.
Pearson Education North Asia Ltd, *Hong Kong*
Pearson Education Canada, Ltd., *Toronto*
Pearson Education de Mexico, S.A. de C.V.
Pearson Education—Japan, *Tokyo*
Pearson Education Malaysia, Pte. Ltd.

CONTENTS

Supplemental Laboratory Topics   299

PREFACE

With the thirteenth edition, *Genetics Laboratory Investigations* is well into its sixth decade serving the needs of genetics instructors and students across the United States and Canada. Beginning in 1952, the first four editions were authored by Eldon Gardner of Utah State University. Thomas Mertens assumed the major responsibility for preparing the next three editions. Robert Hammersmith joined the writing team for the eighth edition; following Dr. Gardner's death, Mertens and Hammersmith prepared editions nine through twelve and now thirteen. This long history of service to the genetics education community was recognized in 1998 by the Text and Academic Authors Association with the presentation of the McGuffey Award.

Much of the success of the book can be accounted for by the periodic inclusion of new coauthors who have brought their special expertise to bear on the content, techniques, and philosophy of the book. Beyond that, the book has been designed to be user-friendly and to meet the needs of both faculty and students, having been developed by those actively involved in the teaching of introductory general genetics at the undergraduate level. We are firmly convinced that the student's laboratory experience in genetics is an essential part of the learning process and vitally important to fostering student interest in, and understanding of, the discipline. Toward this end, a number of "open-ended" investigations have been included in this manual—the most obvious being Investigation 13. With appropriate adaptation, however, instructors can use the open-ended approach in any of the investigations in which original (for the student) data are collected.

A study of the teaching of general genetics in Canadian colleges and universities (T. L. Haffie et al., 2000, *Genome* 43:152–159) revealed that insofar as laboratory work is concerned, there was a heavy emphasis on classical transmission genetics in the introductory course. We believe that this is also true in the general genetics courses taught in the United States. Reviews of our book that were completed by the selection team for the McGuffey Award noted that, while the manual stressed classical genetics, provisions had been made to include recent advances related to molecular and human genetics. A positive review of the twelfth edition in the journal of the Genetics Society of Great Britain (*Heredity* 86:639) agreed with the McGuffey Award selection team. The reviewer, Colin Ferris, of the University of Leicester, did suggest that we needed to include work using online Web resources. This suggestion has led to our citing Web sites whenever that seemed appropriate and to the addition of an entirely new investigation (Investigation 26) utilizing national genomic databases of the National Center for Biotechnology Information (NCBI).

We have tried to live up to the history of the manual in this new edition. As in previous editions, we have included investigations to accommodate courses taught in a variety of academic settings and to be cost-effective, especially for investigations involving molecular genetics . Many investigations provide two or more alternative ways of achieving the instructional objectives. Newly designed line drawings in various investigations and new chromosome spreads for karyotyping in Investigation 10 are among the more visible changes. References at the end of each investigation have been reviewed and updated sources added. Problem sets in Investigations 2, 3 and 22 have been redone. A new

Instructor's Manual provides answers to most questions in each investigation, helpful hints for setting up experiments in certain investigations, and when needed, sources of instructional materials beyond those suggested in the book itself.

The cover of the book is also useful in the instructional process. The photograph of "painted" human chromosomes on the back cover is used to assist students to understand how molecular techniques for differential visualization of chromosomes can be helpful in analyzing complex chromosome rearrangements at the cytogenetic level (see Investigation 11).

To all who have assisted us over the years, we give our thanks. Credit for illustrations and tables in the present edition are given with appropriate investigations according to the wishes of the authors or publishers from whom permissions have been granted.

Investigations 12 (genetics of *Sordaria*), 22 (human fingerprint ridge counts), and 24 (genetic drift) were originally published in *The American Biology Teacher* as was the portion of Investigation 6 dealing with meiosis in *Tradescantia*. We are indebted to Dr. George Hudock of Indiana University for permission to modify two investigations included in his book *Experiments in Modern Genetics* (John Wiley & Sons, 1967). These modifications of Hudock's work appear as our Investigations 18 and 21. We acknowledge with special appreciation new photos of human chromosome spreads (Investigation 10) and the front cover photograph provided by Dr. Daniel L. Van Dyke and his colleagues at the Mayo Clinic. We thank Carolina Biological Supply Company (Investigation 5), the Biological Sciences Curriculum Study (Investigation 22), and Helena Laboratories (Investigation 25) for illustrations they provided. The section of Investigation 25 that deals with electrophoresis of hemoglobin has been reviewed by Helena Laboratories of Beaumont, Texas, and is used with their permission. Investigation 16 is based in part on material supplied by the Perkin Elmer Corporation and by Carolina Biological Supply Company, and thanks are due to both firms for their assistance. Kits for conducting this investigation may be purchased from Carolina Biological. We wish to thank our project manager, Crissy Dudonis, and her staff for their support and efforts in the production of this manual. Finally, we wish to thank the instructors and students who have used the twelfth edition of this manual. It is their commitment to the manual that has justified the production of the thirteenth edition. We hope that the updating and additions found in the new edition will continue to make the manual a useful instructional tool.

T. R. M.
R. L. H.

INVESTIGATION 1

Drosophila and Maize Experiments in Genetics: Monohybrid and Dihybrid Crosses

In 1866, Gregor Mendel published a paper detailing a set of experiments on the common garden pea. From a careful analysis of these experiments, Mendel proposed a set of rules for the inheritance to traits in peas. After the independent rediscovery of Mendel's work by C. Correns, H. de Vries, and E. von Tschermak in 1900, and the subsequent demonstration of the applicability of his rules in many diverse organisms, the principles illuminated by Mendel were renamed the Mendelian laws of inheritance. These two laws of inheritance are referred to as (1) **the law of segregation** and (2) **the law of independent assortment**.

While working on the garden pea, Mendel selected seven distinct traits, each with two different and easily recognizable forms, such as yellow pea versus green pea or tall versus dwarf plants. Each of these variant forms of the same gene (yellow vs. green) we now call alleles of a gene. Mendel found when studying crosses of plants possessing only one of the seven traits (monohybrid crosses), the alleles of a gene (yellow vs. green) segregate from one another when gametes are formed. The law of segregation results from the separation of homologous chromosomes and, thus, alleles during meiosis (Investigation 6) and the subsequent random fertilization of gametes results in the predicted ratios observed in monohybrid crosses (monohybrid F_2 ratios of 3:1 and monohybrid testcross ratios of 1:1).

In considering two traits simultaneously (dihybrid crosses), Mendel postulated independent assortment of the genes. Mendel studied the same seven traits, and in all of the situations investigated, he obtained dihybrid F_2 ratios of 9:3:3:1 and dihybrid testcross ratios of 1:1:1:1.

What is the mechanism of independent assortment? Genes located on separate pairs of chromosomes assort independently of one another. Thus, one might assume that the seven different traits studied by Mendel were controlled by seven gene loci located—one locus per chromosome pair—on the seven different chromosome pairs of the garden pea. This, in fact, is not the case, as was documented by Novitski and Blixt (1978). Because of recombination, genes located on the same chromosome pair, but at great distances from one another, will also assort independently of one another. It is now known that two chromosome pairs in the garden pea each carry more than one of Mendel's seven pairs. In addition, at least two chromosome pairs carry none of Mendel's genes. The independent assortment of Mendel's genes on the same chromosome pair is explained by the great genetic distance between those genes. If, on the other hand, two genes on the same chromosome are relatively close to one another, distortion of dihybrid ratios occurs and the two genes are said to be linked (see Investigation 11).

Investigations with organisms other than the garden pea revealed that independent assortment occurs in them also. Figure 1.6 shows an ear of F_2 kernels of corn; the kernels are distributed according to the law of independent assortment with 9/16 of the kernels being colored with starchy (smooth) endosperm, 3/16 being colored with sugary (wrinkled) endosperm, 3/16 being colorless starchy, and 1/16 being colorless sugary.

Geneticists soon found that dihybrid crosses with independent assortment may not always yield F_2 ratios of 9:3:3:1. If two independently assorting genes both affect the same characteristic—coat color in mice, for example—a modified ratio such as 9 agouti (wild type): 3 black : 4 albino may occur. Such gene interaction is called **epistasis**, and it is a common phenomenon in both plants and animals investigated in genetic studies. Because of epistasis the classical 9:3:3:1 ratio may be modified to produce such variations as 9:3:4, 9:7, 9:6:1, 12:3:1, 13:3, and 15:1.

A number of organisms may be used for demonstrating the classical Mendelian laws. Two eukaryotic organisms that are exceptionally well understood genetically are *Drosophila* and maize (*Zea mays*). In this investigation you will have the opportunity to study both the laws of segregation and independent assortment in both *Drosophila* and maize. In addition, epistasis in the inheritance of aleurone pigmentation in maize endosperm will be investigated.

OBJECTIVES OF THE INVESTIGATION

Upon completion of this investigation, you should be able to

1. **outline** the basic procedures for culturing and experimenting with *Drosophila melanogaster*,
2. **recognize** and **interpret** *Drosophila* and maize F_2 data that illustrate Mendel's Law of Segregation,
3. **recognize** and **interpret** F_2 data that illustrate Mendel's Law of Independent Assortment,
4. **list** the necessary conditions for obtaining independent assortment, and
5. **recognize** and **interpret** dihybrid F_2 data that illustrate genic interactions.

Materials needed for this investigation:

for the class:

morgue containing 70% ethyl alcohol

appropriate stocks of *Drosophila*

F_2 kernels of maize from ears similar to those shown in Figures 1.5 and 1.6

ears of F_2 maize kernels illustrating gene interactions

for each student (Figure 1.1):

stereo dissecting microscope

bottles of culture medium

self-adhering labels

etherizer

re-etherizer (petri dish with absorbent material or filterpaper taped to the inside)

ether in dropping bottle (or other material for anesthetizing flies)

dropper

teasing needles or fine camel's hair brush

The vinegar or fruit fly, *Drosophila melanogaster*, is especially suited to experimental crosses in the classroom laboratory. This small fly passes through a complete metamorphosis in 10 to 14 days at 25°C. In addition to a short life cycle, *Drosophila* possesses an abundance of genetic variability, is highly prolific, and is a convenient and inexpensive organism to study. Before performing genetic experiments with *Drosophila*, however, you must learn some basic facts about the biology and cultivation of this organism.

I. MEDIUM

Numerous media have been developed for the culture of *Drosophila*. Perhaps the easiest to use is "instant medium."[1] One need only add water to the concentrate to produce a medium that is immediately usable.

---

[1] Instant medium is available from Carolina Biological Supply Co., 2700 York Road, Burlington, NC 27215.

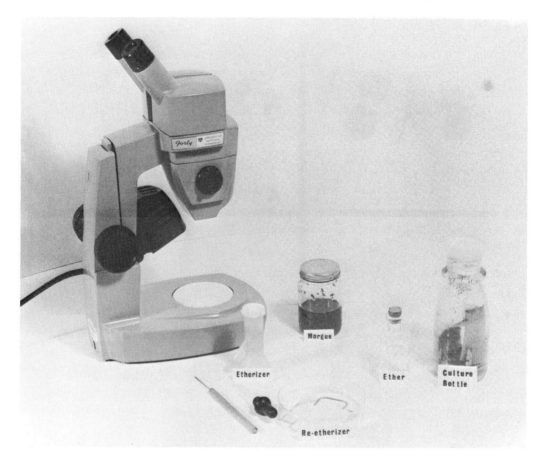

FIGURE 1.1. Equipment needed for handling and examining *Drosophila*.

No cooking is necessary; however, some workers suggest adding five drops of a 5% solution of propionic acid or Dowicil-200[2] (1.5 µg/ml) to assist in inhibiting mold. Although convenient, this medium may prove too expensive when large class enrollments require using a considerable amount of media.

Demerec and Kaufmann's *Drosophila Guide* includes a number of recipes for different media.[3] The ingredients and instructions for making another useful medium, which has been attributed to C.B. Bridges of the California Institute of Technology, also are included in the *Instructor's Manual*.

Your Instructor will provide you with bottles or vials of media or will instruct you in the preparation of the bottles (vials).

II. DEVELOPMENT OF THE FLIES

Drosophila has served as a great experimental organism in large part because of its short, unique life cycle (Figure 1.2). The eggs of *Drosophila* are laid on or near the surface of the culture medium. They hatch about 1 day later, producing minute, white larvae, which burrow in the medium and feed on yeast cells. The larvae attain full size at the end of about 7 days. They then climb onto the side of the bottle and pupate. The pupal stage lasts about 2 days, after which the mature flies emerge. At 25°C, the entire life cycle from egg to adult is completed in about 10 days. Familarize yourself with all of the

---

[2] Dowicil-200 is available from Sigma-Aldrich, P.O. Box 14508, St Louis, 63178.

[3] *Drosophila Guide* may be purchased from Carnegie Institution, Department B, 1530 P Street NW, Washington, DC 20005.

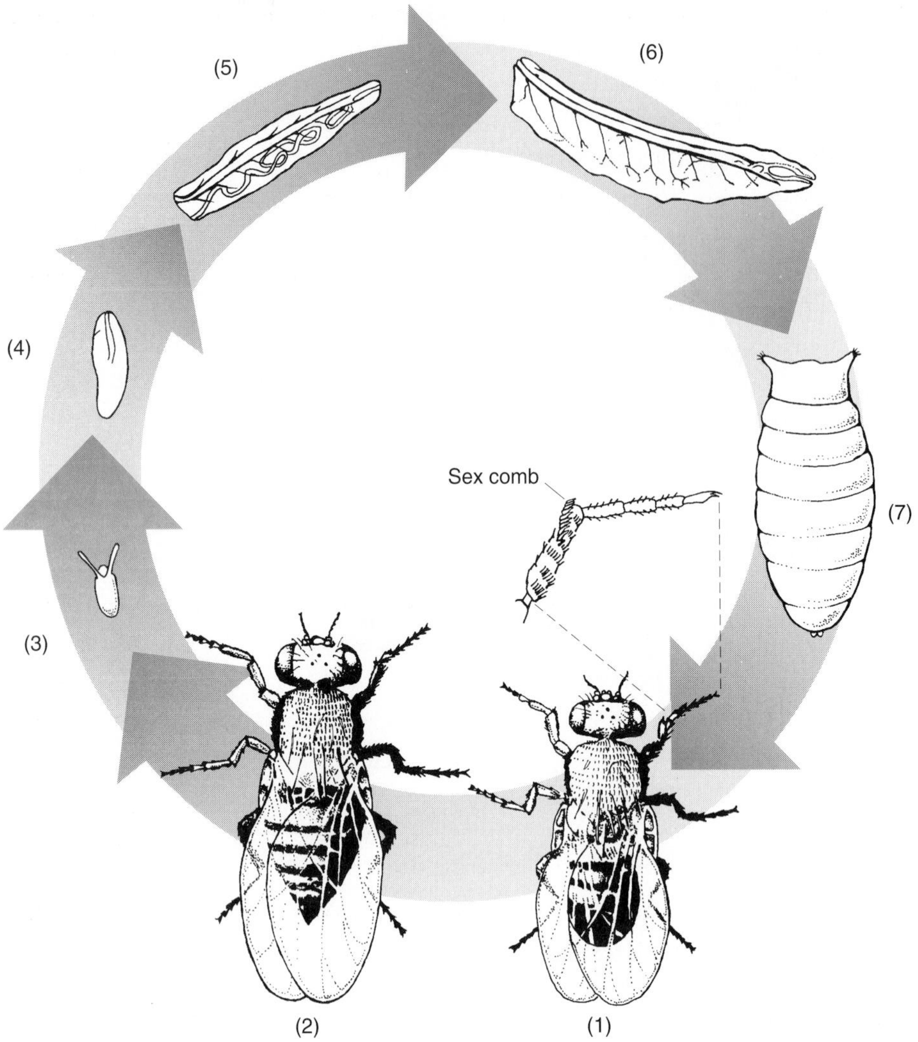

FIGURE 1.2. Life cycle and sexual dimorphism of the vinegar fly, *Drosophila melanogaster.* The male (1) is generally smaller than the female (2). The male also has a more rounded abdomen that is black-tipped rather than striped. The enlarged foreleg of the male reveals special bristles known as the sex comb, a characteristic lacking in the female. Metamorphosis is complete, from egg (3), larval stages (4, 5, and 6), pupa (7), through adult.

stages represented in Figure 1.2. Most of these stages can be seen by simply observing an actively growing bottle of flies.

III. HANDLING THE FLIES

To etherize and examine adult flies, proceed as follows:

1. Place a few drops of ether on the absorbent material of the etherizer.
2. Strike the base of the culture bottle lightly on the palm of the hand so the flies will drop to the bottom.
3. Remove the culture bottle plug, quickly replace it with the mouth of the etherizer, invert the bottle over the etherizer, and shake flies into the etherizer.

4. Subject the flies to ether for about 30 seconds after they cease moving. Avoid overetherization if they are to be used in further matings. The flies will die if left in the etherizer too long. Overetherized flies hold their wings vertically over their body (in contrast to the normal at-rest position shown in Figure 1.2).

5. Transfer the etherized flies to a clean white card. A 3-by-5-in. file card is ideal.

6. Examine the etherized flies with a dissecting microscope at 10× to 25× magnification. Use a soft brush or teasing needle for moving the flies about on the stage of the microscope.

7. If the flies revive before you finish examining them, add a few drops of ether to the absorbent pad on the re-etherizer and cover flies on the microscope stage for a few seconds.

8. If the flies are not needed after observation, they may be immediately discarded in the morgue. Etherized flies to be used for further matings should be permitted to recover in a dry vial or on a dry surface in the culture bottle before they come in contact with the moist medium.

IV. DISTINGUISHING SEX

Examination of the external genitalia under magnification is the best means of distinguishing the sex of flies. Only male flies exhibit darkly colored external genitalia, which are visible on the ventral side of the tip of the abdomen (Figure 1.3). The following characteristics, illustrated in Figures 1.2 and 1.3, can also help to distinguish males from females:

1. **Size.** Females are usually somewhat larger than males.

2. **Shape.** The caudal extremity of the male is round and blunt, whereas that of the female is sharp and protruding. The abdomen of the male is relatively narrow and cylindrical, whereas that of the female is distended and appears spherical or ovate. Adults newly emerged from the pupa case are relatively long and slender, and the sexual differences described here are not so readily noted.

3. **Color.** Black pigment is more extensive on the caudal extremity of the male than on that of the female. On the male, the markings extend completely around the abdomen and meet on the ventral side. On the female, the pigment occurs only in the dorsal region.

4. **Sex combs.** Only males have a small tuft of black bristles called a sex comb on the anterior margin at the basal tarsal joint of each front leg. Magnification is necessary to see the sex combs.

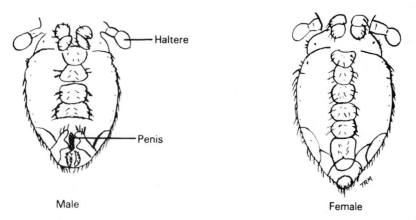

FIGURE 1.3. Ventral view of the abdomens of male and female *Drosophila melanogaster* showing external genitalia.

V. EXPERIMENTAL MATINGS

The methods outlined here are essentially those used in actual research studies with *Drosophila*. Pedigreed stocks carrying mutant genes are maintained in the laboratory. Flies in these cultures tend to breed true (i.e., produce offspring identical in appearance to the parents) as long as they are mated among themselves. Rarely, however, new hereditary variations—spontaneous mutations—do occur.

For classroom purposes, flies having easily recognizable differences are used. Names and symbols used to identify the mutant genes carried by the flies are those devised by research workers. Stocks are conveniently designated by symbols indicating the particular mutant gene or genes carried. Flies that exhibit traits that may be considered standard or normal are designated as wild type. The plus symbol ($+$) indicates wild type with reference to any gene. The lowercase letter indicates that the mutant allele is recessive to the wild-type allele. A capital letter designates a dominant allele. The symbol e, for example, represents the recessive mutant allele for ebony body color and e^+ (also, $+^e$ or E) the dominant allele for wild type, gray body; B symbolizes the dominant mutant allele for bar eye and B^+ ($+^B$ or b) the recessive allele for wild-type eye. Homozygous ebony flies are symbolized ee and homozygous wild-type $e^+ e^+$

A. Becoming Familiar with Fly Characteristics

The first part of this investigation will provide an opportunity to become acquainted with the morphological characteristics of wild-type and mutant flies. Complete Table 1.1, recording with appropriate symbols, sketches, or descriptions the differences you see between mutant and wild-type flies. First examine wild-type flies that are considered normal for all traits listed. Place a plus sign in the appropriate space to signify the normal condition. Carefully examine flies from three or more mutant stocks prepared as unknowns in the laboratory and compare them with wild type. If the unknown flies are wild type for a given trait, place a plus sign in the appropriate space. If they are different from wild type, diagram the mutant trait or use one or two key words to describe it briefly.

B. Securing Virgin Females

To begin a cross between two varieties of flies, you must secure a virgin female. Once inseminated, females retain viable sperm for several days. Thus, the only way to ensure a controlled mating between different genetic stocks is to use virgin females.

The most common method of obtaining virgin females is to select those that have recently emerged from their pupa cases. These flies (*D. melanogaster*) do not mate for at least 8 hours

TABLE 1.1. Comparison of Wild-Type and Mutant *Drosophila*

Trait	Wild Type	Unknown 1	Unknown 2	Unknown 3	Unknown 4
Body color	_____	_____	_____	_____	_____
Eye color	_____	_____	_____	_____	_____
Eye shape	_____	_____	_____	_____	_____
Wing shape and size	_____	_____	_____	_____	_____
Antenna shape and size	_____	_____	_____	_____	_____
Bristle shape and size	_____	_____	_____	_____	_____

(probably not for 10 hours) after emerging from the pupa case. Thus, 8-hour-old females will still be virgins even if male flies are present. To secure virgin females, follow this procedure:

1. Empty the appropriate stock bottle of all adult flies. Record the time on the bottle.
2. Within 8 hours of removing the adult flies, etherize any newly emerged adults. Such newly emerged flies are distinguished by their pale body color and a characteristic dark spot on the ventral side of the abdomen, slightly to the left of the midline.
3. Females among those newly emerged adults will be virgins that can be used in the appropriate cross.

   Another method of obtaining virgin females is to isolate pupae from which the adult flies are about to emerge. Single pupa cases should be placed in small vials containing a narrow strip of moistened paper towel. The vial should be stoppered with cotton. Of necessity, any female hatching alone in a vial will be a virgin. This method of obtaining virgin females is laborious and can prove unwieldy because it requires many vials.

C.   Making Crosses

When making a cross between two varieties, consider only those characters in which the parent flies differ. For example, in crossing flies having ebony body color with those having white eyes, only body color and eye color need to be observed carefully.

   Secure etherized male flies of one variety and etherized virgin females of the other. While holding the culture bottle on its side, place these flies in the bottle. *Be sure to add some dry yeast granules or yeast suspension to the medium before introducing the flies.* Keep the bottle on its side until the flies have recovered from etherization. This position will prevent their becoming stuck in the medium. Label the bottle and record data in Table 1.2. After 7 or 8 days, remove the parent flies (P_1) to prevent their being confused with or mated with their offspring. Flies of the first filial generation (F_1) will soon begin to emerge. After several F_1 flies have appeared, etherize and examine them, especially with regard to the characters by which the P_1 flies differed. Record in Table 1.2 the phenotypes of the F_1 flies of each sex; place these F_1 flies in a fresh bottle of medium. (Be sure to label the bottle!) This mating will allow for the production of a second filial generation (F_2). It is not necessary that the F_1 female flies be virgins for this mating.

1. Why? _____

   The instructor may wish to have you testcross an F_1 female fly. In this case you should use a virgin female.

2. Why is a virgin female needed in this case?_____

   _____

3. What is a testcross?_____

   _____

D.   Laboratory Records

Much of Mendel's success in elucidating the basic principles of inheritance depended on his keeping accurate records about his experiments. Keeping accurate and detailed notes is also necessary to the successful completion of your laboratory work. To this end, you are provided with Table 1.2 in which to keep records as you do the various steps of this investigation. You may want to duplicate this table if your instructor has you performing multiple crosses.

TABLE 1.2. Record of a *Drosophila* Experiment

Record both the phenotype and stock number of the parent flies.

Experiment number _____ **Name** _____

1. Cross _____ female $\times$ _____ male

2. Date P_1s mated: _____

3. Date P_1s removed: _____

4. Date F_1s first appeared: _____

5. Phenotype of F_1 males: _____

6. Phenotype of F_1 females: _____

7. Date F_1 male and female placed in fresh bottle (or date F_1 virgin female testcrossed): _____

8. Date F_1 files removed: _____

9. Date F_2 (or testcross) progeny appeared: _____

10. Record F_2 or testcross data in the following table: _____

	Males		Females		
	Phenotype	Number	Phenotype	Number	Total
a.	_____	_____	_____	_____	_____
b.	_____	_____	_____	_____	_____
c.	_____	_____	_____	_____	_____
d.	_____	_____	_____	_____	_____
e.	_____	_____	_____	_____	_____
f.	_____	_____	_____	_____	_____
g.	_____	_____	_____	_____	_____
h.	_____	_____	_____	_____	_____
Totals		_____		_____	_____

VI. SUGGESTED MONOHYBRID CROSSES

A. *Drosophila* Crosses

A number of crosses might be used to demonstrate Mendel's first law, the **Law of Segregation**. Examples follow:

P_1 Female	$\times$	P_1 Male
1. sepia eye color	$\times$	wild type (red)
2. dumpy wings	$\times$	wild type (long)
3. vestigial wings	$\times$	wild type (long)
4. ebony body color	$\times$	wild type (gray-brown)
5. aristapedia antennae	$\times$	wild type antennae
6. _____	$\times$	_____

   Your instructor may use one of these examples or perhaps the reciprocal crosses of those suggested. For example, the reciprocal of the first mating would consist of mating a wild-type female with a male having sepia eyes. An interesting and instructive experiment would be one comparing the results of such reciprocal matings in both the F_1 and F_2 generations. (Sex-linked traits will exhibit differences in reciprocal crosses; see Investigation 8.)

   The first mating may have been prepared for you 5 or 6 days before today's laboratory period. If so, the bottles you receive will contain P_1 flies and F_1 larvae. In any case, on the basis of the cross that you make (or the one made for you), give the following information:

	Female	Male
Genotype of P_1	_____	_____
Genotype of F_1	_____	_____
Phenotype of F_1	_____	_____
Genotypes of F_2	_____	_____

1. What F_2 phenotypic ratio do you expect to obtain? _____

2. What F_2 genotypic ratio is expected? _____

3. Which trait is dominant in your experiment? _____

4. How do you know? _____

B. Maize Crosses

The inheritance of the extensive genetic variability in maize also obeys Mendelian principles; both seedling and aleurone (endosperm) characters lend themselves to classroom use. Figure 1.4 shows a flat of F_2 seedlings segregating for a recessive mutant allele for albinism (absence of chlorophyll). The classical Mendelian ratio of 3:1 is expected. Similarly, Figure 1.5 shows an ear of F_2 kernels of corn that are segregating in a ratio of 3 colored : 1 colorless aleurone. Can you use the photos to confirm these expections?

FIGURE 1.4. Flat of F_2 corn seedlings segregating for a recessive albino mutation. The classical Mendelian ratio of 3:1 is expected.

Because maize has a relatively long life cycle (3 months or more to complete), you probably will not be able to conduct actual experimental matings with it. Your instructor might provide you with ears of genetic corn or flats of F_2 seedlings and request that you determine the number of individuals having the various phenotypes and then interpret the data in terms of Mendelian principles. For example, suppose you counted 40 green and 12 albino seedlings in an F_2 population. How would you interpret these data?

Ears bearing F_2 kernels can be purchased from a number of biological supply companies. Without removing the kernels from the ears, you can count the number of kernels in the different phenotypes,

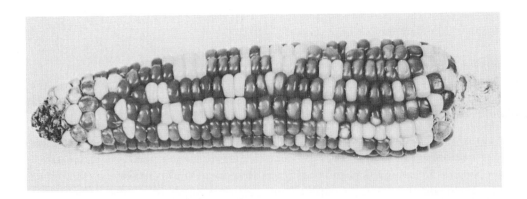

FIGURE 1.5. Ear of F_2 corn kernels segregating for colored versus colorless aleurone. A 3:1 ratio is expected.

TABLE 1.3. Record of F_2 Data for Maize Endosperm Trait

Phenotypes*	Number of Kernels Observed	Number of Kernels Expected	Genotypes of Kernels
_____	_____	_____	_____
_____	_____	_____	_____
Total	_____	_____	

* Phenotypes will vary depending on the trait studied.

record data in Table 1.3, and then formulate hypotheses to explain the data collected. Two possible illustrations of endosperm traits follow:

	Parents		F_1	F_2	
1.	*SuSu*	× *susu*	*Susu*	3 *Su–*	: 1 *susu*
	starchy	sweet	starchy	starchy	sweet
	(smooth)	(wrinkled)	(smooth)	(smooth)	(wrinkled)
2.	*CC*	× *cc*	*Cc*	3 *C–*	: 1*cc*
	pigmented	nonpigmented	pigmented	pigmented	nonpigmented
	(purple)	(white)	(purple)	(purple)	(white)

A host of other endosperm mutants have been researched. Many are suitable for classroom laboratory investigation. Examples include *Wx* (starchy) versus *wx* (waxy), *Y* (yellow) versus *y* (white), and *Pr* (purple) versus *pr* (red) endosperm. Franks (1980, 1981) and Neuffer et al. (1997) provide further information and excellent color photographs of useful maize mutants.

VII. INDEPENDENT ASSORTMENT

A. Dihybrid Cross with *Drosophila*

Under the direction of the instructor, prepare a mating between flies carrying homozygous mutant genes in the second and third chromosomes. Produce F_1 and F_2 generations. Several crosses are appropriate for this experiment. Although one such cross will be suggested here, your instructor may decide to use alternative crosses.

1. The gene for vestigial wings (*vg*) is located on the second chromosome, and the gene for ebony body (*e*) color is located on the third chromosome. Matings may be made between flies from these two stocks. *Be certain to use virgin female flies for this cross.* Reciprocal crosses, that is, vestigial female (*vg vg*; $e^+ e^+$) × ebony male ($vg^+ vg^+$; *ee*) and ebony female × vestigial male, may be made if time and equipment permit. After about 8 days, remove and discard the parent flies. Record all data relative to this experiment in Table 1.4.

2. Predict the results in the F_1 and F_2 generations from the matings prepared.

    F_1 phenotype(s):_____

    F_2 phenotypes and frequencies:_____

    _____

    _____

    _____

3. When the F_1 flies appear in the culture bottles, check their phenotype to see if your expectations have been obtained. Record data relative to the F_1 flies in Table 1.4. Mate F_1 males and F_1 females in new culture bottles. After about 8 days, remove and discard the F_1 flies.

4. When the F_2 adult flies emerge in the culture bottle, etherize and classify them for the various characteristics involved in this cross. Record data in Table 1.4.

TABLE 1.4. Record of a *Drosophila* Experiment

Record both the phenotype and stock number of the parent flies.

Experiment number _____   **Name** _____

1. Cross _____ female × _____ male

2. Date P_1s mated: _____

3. Date P_1s removed: _____

4. Date F_1s first appeared: _____

5. Phenotype of F_1 males: _____

6. Phenotype of F_1 females: _____

7. Date F_1 male and female placed in fresh bottle (or date F_1 virgin female testcrossed): _____

8. Date F_1 files removed: _____

9. Date F_2 (or testcross) progeny appeared: _____

10. Record F_2 or testcross data in the following table: _____

| Males | | Females | | |
Phenotype	Number	Phenotype	Number	Total
a. _____	_____	_____	_____	_____
b. _____	_____	_____	_____	_____
c. _____	_____	_____	_____	_____
d. _____	_____	_____	_____	_____
e. _____	_____	_____	_____	_____
f. _____	_____	_____	_____	_____
g. _____	_____	_____	_____	_____
h. _____	_____	_____	_____	_____
Totals	_____		_____	

TABLE 1.5. Summary of Data and Calculations for the Second *Drosophila* Experiment

Phenotype	Observed Number (O)	Expected Number (E)	Deviation (O − E)
_____	_____	_____	_____
_____	_____	_____	_____
_____	_____	_____	_____
_____	_____	_____	_____
Totals	_____	_____	_____

5. Calculate the expected number of individuals in each of the phenotypic categories, and calculate the deviations (differences) between observed (O) and expected (E) numbers. Deviations are calculated as O − E. Record these data in Table 1.5. These data will be used in Investigation 3.

6. Indicate the genotypes and phenotypes of the flies in each of the three generations.

P_1 _____

F_1 _____

F_2 _____

B. Dihybrid Crosses in Maize

You will be given an envelope of F_2 maize kernels or an intact ear selected by your instructor from one of the alternative crosses listed in the *Instructor's Manual*. These kernels may be taken from an ear comparable to the one shown in Figure 1.6.

1. Classify the kernels into the four different phenotypes. Count and record the number of kernels in each phenotype in the space provided in Table 1.6.

9:3:3:1

FIGURE 1.6. Ear of maize bearing F_2 kernels segregating in the ratio 9:3:3:1.

TABLE 1.6. Summary of Data and Calculations for Dihybrid F_2 in Maize

Calculate the number of kernels expected in each phenotypic category and the deviation between observed and expected (O − E).

Phenotype	Observed Number (O)	Expected Number (E)	Deviation (O − E)
Totals			

2. Select gene symbols to represent the two alleles of each gene studied in this example.

Gene symbol	Phenotype

    a. Which characteristics are dependent on dominant alleles?

    b. Recessive alleles? _____

3. Now, using the gene symbols you have chosen, give the possible genotypes of the original homozygous parents (P_1) of this F_2 generation.

    a. What are the phenotypes of these parent individuals?

    b. Assuming the parents to be homozygous, could more than one original cross have been used to produce this F_2 generation?

       If so, what crosses could have been made? (Give genotypes and phenotypes.)

4. In any case, what must have been the genotype and phenotype of the F_1 generation that subsequently was self-pollinated to produce the F_2 kernels with which you have been working?

_____

_____

VIII. GENE INTERACTIONS IN MAIZE

Numerous gene loci scattered over the 10 pairs of chromosomes of maize determine endosperm color as well as other endosperm characteristics. In this exercise you will determine how some of these genes interact to determine endosperm color.

You will be given an ear of corn containing F_2 kernels of two or more colors.[1] Count the number of F_2 kernels of each color. From the F_2 data and from the information provided about the P_1 and F_1 phenotypes, determine the mechanisms of inheritance that account for the data.

Among the possible kinds of ears of corn your instructor may ask you to study are the following:

	Phenotypes of the true-breeding parents	Phenotype of the F_1 kernels	Phenotypes of the F_2 kernels
1.	purple × white	purple	purple, red, white
2.	yellow × yellow	yellow	yellow, purple
3.	white × white	purple	purple, white
4.	purple × yellow	purple	purple, yellow, white
5.	_____	_____	_____

Your instructor will give you an ear of corn and tell you which of the four kinds it is, or if it is a special mating selected by your instructor.

1. Record the number of the ear here: _____

2. Record the phenotypes of the true-breeding parental generation here:

_____

3. Record the phenotype of the F_1 kernels here: _____

_____

4. Count the F_2 kernels on the ear and record their phenotypes, numbers, and ratios in the following spaces:

Phenotype	Number	Ratio suggested by F_2 data
_____	_____	_____
_____	_____	_____
_____	_____	_____
Total	_____	

---

[1] Ears of F_2 corn needed for this investigation are available from Carolina Biological Supply Co., 2700 York Road, Burlington, NC 27215.

5. How do genes interact to produce such epistasis ratios as 9:3:4? The following account explains the inheritance of mouse coat color: A true-breeding agouti (wild-type) mouse has the genotype *CCAA*, and a true-breeding albino may have the genotype *ccaa*. If mated, two such animals would produce F_1 offspring having the genotype *CcAa* and the agouti phenotype. In the F_2 generation produced from crossing the F_1's among themselves, one would expect 9 *C–A–* : 3 *C–aa*: 3 *ccA–*: 1 *ccaa*. The 9/16 that are *C–A–* will be agouti, and the 1/16 that is *ccaa* will be albino, but what about the 3/16 *C–aa* and the 3/16 *ccA–*? Those having the *ccA–*phenotype will also be albino because they are homozygous recessive *cc*, a genotype that prevents them from producing any pigment whatsoever. Those individuals constituting the 3/16 *C–aa* will be able to make pigment because they have the dominant *C* allele, but the pigment that they produce will be black because they are homozygous recessive *aa*. In summary, 9/16 *C–A–*can make agouti pigment, 3/16 *C–aa* can make black pigment, and 4/16 (= 3/16 *ccA–*plus 1/16 *ccaa*) can make no pigment at all and are thus albinos. Now attempt to devise a hypothesis to explain the mechanism of inheritance of endosperm color for the ear of corn you have been studying. In doing so, answer the following questions:

a. What ratio do the F_2 data approximate? _____

_____

b. How many gene loci appear to determine the F_2 phenotypes?

_____

_____

c. Suggest gene symbols to be used in explaining the data. Indicate how each gene and its allele is functioning to regulate endosperm color by completing the following table:

Gene symbol	**Gene effect**
_____	_____
_____	_____
_____	_____
_____	_____

d. Now, using the gene symbols you have devised, write the following genotypes:

Parental generation	**Phenotype**	**Genotype**
Parent no. 1	_____	_____
Parent no. 2	_____	_____
F_1 generation:	**Phenotype**	**Genotype**
	_____	_____
F_2 generation:	**Phenotypes**	**Genotypes**
	_____	_____
	_____	_____
	_____	_____

e. You have already suggested a ratio based on your F_2 data. How closely do the data approximate the ratio you have hypothesized? To answer the question, calculate the expected numbers (E) for each F_2 phenotype based on the ratio you have hypothesized and the total number of F_2 kernels you have counted. Record your calculations in the following spaces:

F_2 phenotypes	Observed nos. (O)	Expected nos. (E)
_____	_____	_____
_____	_____	_____
_____	_____	_____
Totals	_____	_____

If the expected numbers closely approximate the observed numbers, we conclude that the data seem to be satisfactorily explained by the proposed hypothesis. In Investigation 3 you will study a statistical procedure, the chi-square test, that can be used to determine whether data you have collected are a reasonable approximation of the numbers expected on the basis of a particular hypothesized ratio.

REFERENCES

ASHBURNER, M., K.G. GOLIC, and R.S. HAWLEY. 2005. *Drosophila: a laboratory handbook*, 2nd ed. Cold Spring Harbor, NY: Cold Spring Harbor Laboratory Press.

CORRIHER, C.M. 1985. Modern *Drosophila* care. *Carolina Tips* 48(2):5–6.

DEMEREC, M., and B.P. KAUFMANN. 1986. *Drosophila guide*, 9th ed. Washington, DC: Carnegie Institution of Washington.

FLAGG, R.O. 1979. *Carolina Drosophila manual.* Burlington, NC: Carolina Biological Supply Co.

FORD, R.H. 2000. Inheritance of kernel color in corn: explanations and investigations. *The American Biology Teacher* 62(3): 181–188.

FRANKS, R.L. 1980. Growing genetic corn. *Carolina Tips* 43(7):29–32.

———. 1981. Corn for teaching genetics. *Carolina Tips* 44(5):21–23.

GREENSPAN, R.J. 2004. *Fly pushing: the theory and practice of Drosophila genetics.* Cold Spring Harbor, NY: Cold Spring Harbor Laboratory Press.

GRIFFITHS, A.J.F., ed. 2000. *An introduction to genetic analysis*, 7th ed. New York: W. H. Freeman and Co.

HENDERSON, D.S. 2004. *Drosophila* cytogenetics protocols. *Methods in molecular biology.* vol. 247. Totowa, NJ: Humana Press Inc.

KLUG, W.S., and M.R. CUMMINGS. 2006. *Concepts of genetics.* 8th ed. Upper Saddle River, NJ: Prentice Hall.

KOHLER R.E. 1994. *Lords of the fly: Drosophila genetics and the experimental life.* Chicago: The University of Chicago Press.

LINDSLEY, D.L., and E.H. GRELL. 1968. *Genetic variations of Drosophila melanogaster.* Washington, DC: Carnegie Institution of Washington.

LINDSLEY, D.L., and G.G. ZIMM. 1992a. *The genome of Drosophila melanogaster.* San Diego, CA: Academic Press.

———. 1992b. *The genome of Drosophila melanogaster/chromosome maps.* San Diego, CA: Academic Press.

MERTENS, T.R. 1971. On teaching meiosis and Mendelism. *The American Biology Teacher* 33(7):430–431.

———. 1971. Additional notes on *Drosophila. Carolina Tips* 34(8):29–30.

———. 1971. Teaching Mendel's second law. *Journal of Heredity* 62(1):48–52.

———. 1985. The birthplace of genetics: a historical note. *Journal of Heredity* 76(1):67–68.

NEUFFER, M.G., E.H. COE, and S.R. WESSLER. 1997. *Mutants of maize*, 2nd ed. Plainview, NY: Cold Spring Harbor Laboratory Press.

NOVITSKI, E., and S. BLIXT. 1978. Mendel, linkage, and synteny. *BioScience* 28(1):34–35.

STRICKBERGER, M.W. 1962. *Experiments in genetics with Drosophila.* New York: John Wiley & Sons, Inc.

SULLIVAN, W., M. ASHBURNER, and R.S. HAWLEY. 2000. *Drosophila protocols.* Cold Spring Harbor, NY: Cold Spring Harbor Laboratory Press.

INVESTIGATION 2

Principles of Probability

Fundamental probability principles underlying Mendel's basic laws of heredity will be considered in today's laboratory work. Much of this work will be problem solving. You will investigate the probability of independent events occurring simultaneously and the probability of either one or the other of two mutually exclusive events occurring. You will learn to expand the binomial $(a + b)^n$ to calculate probability for certain combinations of events. Finally, you will use probability principles as a genetic counselor might use them to advise parents of the chance that they will have a child with a genetic abnormality and thereby help them make informed procreative decisions. To this end, you will be asked to study a pedigree and to determine the probabilities of producing an offspring having a genetic defect from certain matings within the pedigree.

OBJECTIVES OF THE INVESTIGATION

Upon completion of this investigation, you should be able to

1. **define** and **give** an example of what is meant by the concept of chance,
2. **use** probability principles in solving problems concerning
   - independent events occurring simultaneously,
   - binomial expansion, and
   - mutually exclusive events, and
3. **apply** probability concepts to the analysis of human pedigrees and to the prediction that certain consanguineous marriages will be at risk for producing an offspring with a genetic defect.

I. THE IDEA OF CHANCE

First of all, consider a simple demonstration of the operation of chance (i.e., probability) in the tossing of coins. It is usually impossible to measure or control the many factors involved when a coin is tossed and allowed to return to the resting position. The end result is said to occur by **chance**. The coin returns to a resting position with one or the other of its faces upturned, that is, heads or tails. When a single coin is tossed many times, it is expected that about half of the tosses will result in heads facing up and the other half in tails facing up.

1. Toss a single coin 24 times. Record the results in Table 2.1. Calculate the expected number of heads and tails and determine the deviations (O − E) between observed and expected. Be sure to indicate whether each deviation is a positive or a negative number. What is the sum of the deviations?

TABLE 2.1. Results of Tossing a Coin 24 Times

Results	Observed Number (O)	Expected Number (E)	Deviation (O − E)
Heads (H)	_____	_____	_____
Tails (T)	_____	_____	_____
Totals	24	_____	0

2. If the deviations (O − E) obtained are small, you can attribute them to chance. If they are large, you must attribute them to some cause other than chance. After you study the chi-square test in Investigation 3, you may wish to return to the data in Table 2.1 to decide whether the deviations are too large to be attributed reasonably to chance alone.

3. What biological and genetic situations are analogous to, and result in, the 50:50 expectation one observes when flipping a coin? The following questions indicate a few biological parallels to flipping a coin: What is the probability that the child expected in the Smith family will be a boy? A girl? If you randomly select 100 families having only one child, in how many of these families might you expect the child to be a boy? A girl? What is the probability that, if the mating $Aa \times aa$ is made, the offspring will have the genotype Aa? Genotype aa?

II. INDEPENDENT EVENTS OCCURRING SIMULTANEOUSLY

Now toss two coins together 36 times. Record the results in Table 2.2. In calculating the expected number in each category, keep in mind that, when two or more coins are tossed together, each is independent and has an equal chance of falling heads or tails. The results obtained from a series of tosses vary with the number of coins tossed together. The expected results can be postulated on the basis of the following generalization: *The probability of two or more independent events occurring simultaneously is the product of their individual probabilities.*

1. If, for example, two coins are tossed together, then the chance for each falling with heads up is 1/2. Likewise, the chance for tails is 1/2. Therefore, the chance that *both* will be heads is

_____. The chance for the first coin to fall heads and the second coin tails is also

TABLE 2.2. Results of Tossing Two Coins 36 Times

Results	Combinations	Observed Number (O)	Expected Number (E)	Deviation (O − E)
Heads on both coins	HH	_____	_____	_____
Heads on one, tails on the other coin	HT, TH	_____	_____	_____
Tails on both coins	TT	_____	_____	_____
Totals	4	36	_____	_____

_____, but one head and one tail can be obtained in another way: The first coin could be tails and the second heads. Therefore, the *total* probability of obtaining a head on one coin and a tail on the other is _____. The chance for both coins to fall tails up is the same as for both to be heads, that is, _____. To summarize, two coins tossed together many times are expected to fall heads, heads about _____ of the time; heads, tails (and vice versa) about _____ of the time; and tails, tails about _____ of the time. Stated as a *ratio* instead of as fractions, the expected result is _____.

2. Note that this situation is similar to what one sees in a simple monohybrid cross. When *Aa* produces gametes, the probability is that 1/2 of the gametes will contain the *A* allele and 1/2 will contain *a*. When an *Aa* female is crossed with an *Aa* male and progeny are produced, the probability is 1/4 that an *A* egg and an *A* sperm will come together to produce an *AA* offspring. Similarly, the probability is 1/2 for *Aa* and 1/4 for *aa* progeny. In studying this monohybrid cross, you are considering a situation in which independent events (the union of different kinds of gametes) occur simultaneously. Thus, a basic probability principle underlies Mendel's first law.

3. The same law of probability applied to the experiment with two coins can be used to calculate the expected ratios when three, four, or more coins are tossed together.
   a. For example, calculate the expected results from tossing three coins together. Let H represent the individual coins that fall heads and let T indicate tails for the same coins. Complete Table 2.3 showing the various possible combinations and the chances of their occurring. The first combination of three heads, which can occur only one way, that is, when all three coins show heads, is worked out as an example. Finally, toss three coins 64 times and record the frequency of occurrence of each of the classes. Calculate the expected numbers and deviations.
   b. A parallel situation would be observed were you to study families consisting of three children. If you were to select randomly 160 families each of which had three children, in how many would the three children be expected to be all boys? Two boys and a girl? One boy and two girls? Three girls?

TABLE 2.3. Expected Results from Tossing Three Coins Together

Write in the letters (H and T) to represent the coins in the "Combinations" column, and calculate the probability for each class resulting when three coins are tossed. Toss three coins 64 times to obtain the observed numbers.

Classes	Combinations	Probability of Each Class Occurring	Observed Number (O)	Expected Number (E)	Deviation (O − E)
3 heads	HHH	1/2 × 1/2 × 1/2 = 1/8	_____	8	_____
2 heads: 1 tail	HHT, HTH, THH	_____	_____	_____	_____
1 head: 2 tails	_____	_____	_____	_____	_____
3 tails	_____	_____	_____	_____	_____
Totals	_____	_____	64	_____	_____

TABLE 2.4. Expected Results from Tossing Four Coins Together

Write in the letters (H and T) to represent the coins in the "Combinations" column, and calculate the probability for each class resulting when four coins are tossed.

Classes	Combinations	Probability of Each Class Occurring
4 heads	HHHH	_____
3 heads : 1 tail	HHHT, HHTH, HTHH, THHH	_____
2 heads : 2 tails	_____	_____
1 head : 3 tails	_____	_____
4 tails	_____	_____

4. Now, calculate the expected results for tossing four coins together. Complete Table 2.4, showing the probability of the different combinations.

5. Assuming that the probability that a boy will be born is 1/2 and that the chance for a girl is also 1/2, answer the following questions.

   a. If four babies are born in a given hospital on the same day, what is the probability that all four will be boys? _____

   b. What is the probability that three will be boys and one a girl?

   _____

   c. What is the probability that two will be boys and two girls?

   _____

   d. What combination of boys and girls among the four babies is most likely to occur?

   _____

   Why? _____

   e. What is the probability that if a fifth child is born it will be a boy?

   _____

   A girl? _____

III. BINOMIAL EXPANSION

In Sections I and II-1 you determined combinations empirically by actually tossing coins (first a single coin and then two coins) and observing the results. Section II-1 discussed a generalization that could be carried over to larger numbers of coins (Sections II-3 and II-4). In Section II-3 you obtained additional empirical data to compare with expected numbers calculated on the basis of the principle first developed in Section II-1. Finally, you saw that if a 1:1 ratio, similar to that for heads and tails on coins, is assumed for the birth of boys and girls in human populations, the same combinations would be expected for babies as for coins (compare Section II-4 with Section II-5).

Expectations for various combinations in groups of given size (n) can be obtained mathematically. Mendel and others recognized that combinations can be calculated by expanding the binomial $(a + b)^n$, in which n is the size of the group, a is the probability of the first event, b is the probability of the alternative event, and $a + b = 1$. In the example involving babies, a = probability of girls = 1/2 and b = probability of boys = 1/2. Now, if you wish to consider the case in which four babies are born in the hospital in 1 day, expand $(a + b)^4 = a^4 + 4a^3b + 6a^2b^2 + 4ab^3 + b^4$.

Expectation of 4 girls = $a^4 = (1/2)^4 = 1/16$
Expectation of 3 girls, 1 boy = $4a^3b = 4(1/2)^3(1/2) = 4/16 = 1/4$
Expectation of 2 girls, 2 boys = $6a^2b^2 = 6(1/2)^2(1/2)^2 = 6/16 = 3/8$
Expectation of 1 girl, 3 boys = $4ab^3 = 4(1/2)(1/2)^3 = 4/16 = 1/4$
Expectation of 4 boys = $b^4 = (1/2)^4 = 1/16$

1. If five babies are born in a given hospital on the same day, what is the chance of (a) 5 boys; (b) 4 boys, 1 girl; (c) 3 boys, 2 girls; (d) 2 boys, 3 girls; (e) 1 boy, 4 girls; (f) 5 girls? To answer this question, you expand the binomial $(a + b)^5 = a^5 + 5a^4b + 10a^3b^2 + 10a^2b^3 + 5ab^4 + b^5$ and substitute $a = 1/2$ and $b = 1/2$ in each term. Do these steps, and answer each part of the question in the following spaces.

(a) _____   (b) _____   (c) _____

(d) _____   (e) _____   (f) _____

When the probability of only a certain combination in a given-sized group is required, factorials may be employed, as indicated in the following formula:

$$P = \frac{n!}{x!(n - x)!} p^x q^{n-x}$$

where P is the probability to be calculated, $n!$ is the product of the number of integers in the group of items $[n! = n \times (n - 1) \times (n - 2) \times (n - 3) \times \cdots \times 1$; that is, if a group consists of five items (such as the five babies in Section III-1), $n! = 5! = 5 \times 4 \times 3 \times 2 \times 1 = 120]$, $x!$ is the product of the integers for one class (p) and $(n - x)!$ is the product of the integers in the other class (q), and p is the probability for one occurrence (i.e., boys) and q is the probability for the other (i.e., girls). *Note:* Factorial 0 $(0!) = 1$ and any number raised to the 0 power = 1.

2. If six babies are born in a given hospital on the same day, what is the probability that two will be boys and four will be girls? Substitute in the following formula:

$$P = \frac{n!}{x!(n - x)!} p^x q^{n-x} = \frac{6!}{2!4!} (1/2)^2(1/2)^4$$

$$= \frac{6 \times 5 \times 4 \times 3 \times 2 \times 1}{2 \times 1(4 \times 3 \times 2 \times 1)} (1/2)^2(1/2)^4 = 15 \times 1/4 \times 1/16 = 15/64$$

Now, using the same procedure, calculate the probabilities that, among six children born in a given hospital on a particular day,

a. one will be a boy and five will be girls. _____

b. three will be boys and three will be girls. _____

c. all six will be girls. _____

3. Albinism in humans is controlled by a recessive gene (*c*). From marriages between two normally

   pigmented carriers (*Cc*), what is the probability of having a normal child? _____. An albino?

   _____. Assume that from such a marriage (*Cc* × *Cc*) four children will be produced.

   a. What is the probability that all four will be normal? _____

   b. That three will be normal and one albino?[1]_____

   c. Two normal and two albino? _____

   d. One normal and three albino? _____

   e. All four albino? _____

IV. EITHER-OR SITUATIONS (MUTUALLY EXCLUSIVE EVENTS)

An additional probability principle is useful in solving certain problems: *The probability of either one or the other of two mutually exclusive events occurring is the sum of their individual probabilities.* Two examples will illustrate how this principle has implications for genetic studies.

1. What is the probability that an individual with the genotype *Cc* will produce either *C* or *c*

   gametes? _____ Obviously the answer must be 1 (i.e., 100%); that is, *Cc* can produce only two kinds of gametes, those containing *C* and those containing *c*. On the basis of the principle just presented, 1/2 = probability that the gamete will contain *C* and 1/2 = probability that the gamete will contain *c*. Therefore, the probability that the gamete will contain either *C* or *c* is 1/2 + 1/2 = 1.

2. If *Aa* is mated to *Aa*, what is the probability that the offspring will have either the genotype *AA*

   or the genotype *Aa*? _____
   The probability for *AA* = 1/4; that for *Aa* = 1/2. Therefore, 1/4 + 1/2 = 3/4 = probability of either *AA* or *Aa*.

The following example illustrates how both principles of probability can be combined to solve problems in genetics.

Probability can also be a useful tool in predicting the results of dihybrid and trihybrid matings. For example, in mating *AaBb* × *AaBb*, one would expect 1/4 of the progeny to be *AA*, 2/4 *Aa*, and 1/4 *aa*, or 3/4 *A*– to 1/4 *aa*. Likewise, 1/4 should be *BB*, 2/4 *Bb*, and 1/4 *bb*, or 3/4 *B*– to 1/4 *bb*. Now, if the *A*(*a*) and *B*(*b*) genes are independently assorting, one would expect 3/4 × 3/4 = 9/16 of the progeny to be *A–B–*, 1/4 × 1/4 to be *AAbb*, and so forth. Use this approach in answering the following series of questions. If *AaBb* is mated to *AaBb*, what is the probability that the offspring will have

---

[1] Note that the theoretical expectation from the cross *Cc* × *Cc* is a 3:1 ratio. However, when only four progeny are produced, the probability is actually *less than one half* (27/64) that the expected 3:1 ratio will be obtained! Thus, geneticists generally favor experimental organisms that produce many progeny so that theoretically expected ratios can be more readily approximated.

a. either the phenotype *A–bb* or the genotype *aaBB?* _____

b. either the genotype *AABB* or the genotype *AaBb?* _____

c. either the phenotype *AaB–* or the genotype *AAbb?* _____

d. either the phenotype *A–B–* or the genotype *aaBB?* _____

V. PROBABILITY AND GENETIC COUNSELING

In many genetic counseling situations, the counselor will prepare a pedigree for the family or families seeking advice. Insofar as possible, the counselor will determine the phenotype and the genotype for each person in the pedigree with respect to the trait in question (e.g., albinism). The counselor can then apply probability principles to determine the probability that a child having a particular abnormality will be produced among the offspring of a certain marriage.

To illustrate the application of these concepts, consider the pedigree shown in Figure 2.1. (For the purposes of this problem, your instructor will indicate which members of the pedigree express the genetic abnormality.) Unless there is evidence to the contrary, assume that individuals who have married into this family do *not* carry the recessive gene for the trait.

For a specific example, you might assume that four members of this pedigree are albinos: the woman in the first generation, her second daughter (the mother of individuals 5, 6, 7, and 8), and individuals 4 and 11 in the third generation. Now, assume that you are a genetic counselor and that individuals 6 and 12 in the third generation of this pedigree come to you and ask, "What is the probability that if we marry and have a family, an albino child will be born to us?"

The counselor must determine the probability that individuals 6 and 12 are heterozygous carriers of the recessive gene for albinism. The counselor must also consider the probability of two heterozygous carriers producing a homozygous recessive child. First of all, the mother of individual 6 is an albino (*cc*), which means that 6 is of necessity (probability = 1) a heterozygote. The father of individual 12 must be heterozygous (*Cc*) since *his* mother is an albino. Although individual 12 is not an albino, he has a 1/2 chance of being heterozygous, depending on which allele (*C* or *c*) he inherited from his father. Finally, two heterozygotes (*Cc* × *Cc*) have 1 chance in 4 of producing an albino (*cc*) child.

The genetic counselor may advise individuals 6 and 12 that if they were to have children, the probability of their having an albino child will be $1 \times 1/2 \times 1/4 = 1/8$. This is the probability that the three independent events will occur simultaneously.

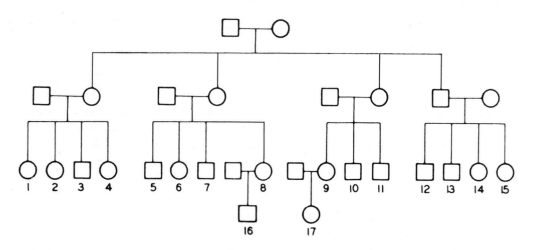

FIGURE 2.1. Human pedigree showing four generations. Circles represent females; squares represent males.

A slightly more confusing situation arises in determining the probability that an unaffected child of two known-heterozygous parents is itself heterozygous. For example, in the pedigree shown, we know with certainty that the parents of individual 10 must be heterozygous. The question, then, is "What is the probability that one of their phenotypically normal children is heterozygous?" We know that in the mating $Cc \times Cc$ the expected ratio of offspring is $1CC:2Cc:1cc$, that is, among the expected three normal ($1CC + 2Cc$) offspring, two may be expected to be heterozygous. Thus, the probability is 2/3 that the normal offspring of heterozygous parents will themselves be heterozygous.

Continue to assume that the same four individuals in the pedigree are albinos, and calculate the probability of the albino trait appearing in the offspring if the following cousins and second cousins should marry and reproduce (show your calculations):

a. 4×5 _____

b. 6×10 _____

c. 7×14 _____

d. 2×10 _____

e. 16×15 _____

f. 16×17 _____

g. 2×12 _____

REFERENCES

BISHOP, O.N. 1984. *Statistics for biology: a practical guide for the experimental biologist*, 4th ed. London: Longman Group, Ltd.

DeGROOT, M.H., and M.J. Schervish. 2001. *Probability and statistics*, 3rd ed. Boston: Addison Wesley.

EVERITT, B.S. 1999. *Chance rules: an informal guide to probability, risk and statistics*. New York: Springer-Verlag.

GRIMMETT, G.R., and D.R. Stirzaker. 2001. *Probability and random processes*, 3rd. ed. New York: Oxford University Press.

GOLDBERG, S. 1987. *Probability: an introduction*. Mineola, NY: Dover Publications.

KOOSIS, D.J., and A.P. Coladarci. 1973. *Probability*. Self-teaching guide. New York: John Wiley & Sons.

———. 1997. *Statistics*, 4th ed. Self-teaching guide. New York: John Wiley & Sons.

ROSS, S. 2005. *First course in probability*, 7th ed. Upper Saddle River, NJ: Prentice Hall.

STIRZAKER, D. 1999. *Probability and random variables: a beginner's guide*. New York: Cambridge University Press.

———. 2003. *Elementary probability*, 2nd ed. New York: Cambridge University Press.

STONE, C.J. 1996. *A course in probability and statistics*. Pacific Grove, CA: Duxbury Press.

STRAIT, P.T. 1988. *A first course in probability and statistics with applications*, 2nd ed., New York: Harcourt Brace Jovanovich.

Note: All references cited in Investigation 3 are also pertinent to this investigation.

INVESTIGATION 3

The Chi-Square Test

The purpose of the chi-square (χ^2) test is to determine whether experimentally obtained data constitute a good fit to, or a satisfactory approximation of, a theoretical, expected ratio; that is, the χ^2 test enables one to determine whether it is reasonable to attribute deviations from a perfect fit to chance. Obviously, if deviations are small, they can be more reasonably attributed to chance than if they are large. The question we try to answer with the χ^2 test is "How small must the deviations be to be attributed to chance alone?" The formula for χ^2 is as follows:

$$\chi^2 = \Sigma \frac{(O - E)^2}{E}$$

where O = the observed number of individuals in a particular phenotype, E = the expected number in that phenotype, and Σ = the summation of all possible values of $(O - E^2)/E$ for the various phenotypic categories.

Before you apply the χ^2 test to any of the data you have collected in Investigations 1 and 2, use the following example for practice: In a cross of tall tomato plants to dwarf ones, the F_1 consisted entirely of tall plants and the F_2 consisted of 102 tall and 44 dwarf plants. Do these F_2 data fit a ratio of 3:1? To answer this question, χ^2 was calculated; the calculations are summarized in Table 3.1.

The calculated χ^2 value is 2.0548, but what does this χ^2 value mean? If the observed numbers (O) were exactly equal to the corresponding theoretically expected numbers (E), the fit would be perfect, and χ^2 would be zero. Thus, a small value for χ^2 indicates that the observed and expected ratios are in close agreement, whereas a large value indicates marked deviation from the expected ratio. Because chance deviations from the theoretical values are expected to occur, the question to be answered is "Are the observed deviations within the limits expected by chance?" Generally, statisticians have agreed on the arbitrary limits of odds of 1 chance in 20 (probability = .05) for drawing the line between acceptance and rejection of the hypothesis as a satisfactory explanation of the data tested.

A χ^2 value of 3.841 for a two-term ratio corresponds to a probability of 1 in 20, or .05. One would obtain a χ^2 value of 3.841 resulting from chance alone in only about 5% of similar trials if the hypothesis is correct. When χ^2 exceeds 3.841 for a two-term ratio, the probability that the deviations can be

TABLE 3.1. Summary of the Calculation of χ^2 in a Hypothetical Illustration

Phenotype	Genotype	O	E	O − E	$(O - E)^2$	$(O - E)^2/E$
Tall	$T–$	102	109.5	−7.5	56.25	0.5137
Dwarf	tt	44	36.5	7.5	56.25	1.5411
Totals		146	146.0			$\chi^2 = 2.0548$

TABLE 3.2. Table of χ^2 Values[*]

df	P = .99	.95	.80	.50	.20	.05	.01
1	.000157	.00393	.0642	.455	1.642	3.841	6.635
2	.0201	.103	.446	1.386	3.219	5.991	9.210
3	.115	.352	1.005	2.366	4.642	7.815	11.345
4	.297	.711	1.649	3.357	5.989	9.488	13.277
5	.554	1.145	2.343	4.351	7.289	11.070	15.086
6	.872	1.635	3.070	5.348	8.558	12.592	16.812
7	1.239	2.167	3.822	6.346	9.803	14.067	18.475
8	1.646	2.733	4.594	7.344	11.030	15.507	20.090
9	2.088	3.325	5.380	8.343	12.242	16.919	21.666
10	2.558	3.940	6.179	9.342	13.442	18.307	23.209
15	5.229	7.261	10.307	14.339	19.311	24.996	30.578
20	8.260	10.851	14.578	19.337	25.038	31.410	37.566
25	11.524	14.611	18.940	24.337	30.675	37.652	44.314
30	14.953	18.493	23.364	29.336	36.250	43.773	50.892

[*] Selected data from R.A. Fisher and F. Yates, *Statistical tables for biological, agricultural and medical research* (London: Oliver and Boyd, 1943).

attributed to chance alone is less than 1 in 20. The hypothesis of compatibility between the observed and expected ratios is thus rejected. In the practice example, $\chi^2 = 2.0548$, which is considerably less than 3.841, the maximum allowable value for a two-term ratio associated with the probability of 1 in 20. Therefore, you can attribute the deviations to chance alone and accept the hypothesis that the data fit a 3:1 ratio.

Where did the value of 3.841 come from? Statisticians have published extensive tables listing χ^2 values (see Table 3.2). Notice that across the top of Table 3.2 are probability (P) values; down the left side are "degrees of freedom" values ($df = 1, 2, \ldots, 30$). The number of degrees of freedom is generally one less than the number of terms in the ratio. In the practice example, with two terms in the ratio (3:1), you have one degree of freedom in interpreting χ^2. Thus, in the one-degree-of-freedom row and under the .05 column you find the χ^2 value of 3.841, the maximum value of χ^2 that we are willing to accept and still attribute the deviations observed to chance alone.

In the hypothetical example, χ^2 was calculated to be 2.0548. Looking in the one-degree-of-freedom row, you see that this value of χ^2 falls between the .05 ($\chi^2 = 3.841$) and the .20 ($\chi^2 = 1.642$) probability columns. This means that the probability that the deviations observed may be attributed to chance alone is between 5% and 20%; that is, if you were to do this same experiment 100 times, you would expect to observe deviations as large as you obtained or larger in between 5 and 20 of the 100 trials owing to chance alone. With this background in the meaning, calculation, and interpretation of χ^2, you can apply the test to some data you have collected yourself.

OBJECTIVES OF THE INVESTIGATION

Upon completion of this investigation, you should be able to

1. **calculate** χ^2 to determine whether a given set of data approximates a theoretically expected ratio and
2. **interpret** a calculated χ^2 value, given the appropriate number of degrees of freedom and a table of χ^2 values.

Materials needed for this investigation:
  container of equal quantities of colored and white beans
  calculator

I. APPLICATION OF THE CHI-SQUARE TEST

A. Application of Chi-Square to New Data

Colored and white beans were carefully selected for equality of size and uniformity of shape. *Equal* quantities of each color were thoroughly mixed and placed in a single container.

Remove a random sample (one petri dish cover or plastic cup full) of this mixture. Such a sample is shown in Figure 3.1. Segregate and count the beans of the different colors. Record your data in Table 3.3, and then calculate the expected numbers based on the size of the sample and the known ratio of colored to white beans in the entire population. Complete Table 3.3 and calculate χ^2.

FIGURE 3.1. Sample from a large container of equal quantities of colored and white beans. Petri dish lid that held the sample is shown adjacent to beans.

TABLE 3.3. Calculation of χ^2 for a Sample Removed from a Large Population Consisting of Equal Numbers of Colored and White Beans

Classes (Phenotypes)	Observed (O)	Expected (E)	Deviations (O − E)	$(O - E)^2$	$(O - E)^2/E$
Colored	_____	_____	_____	_____	_____
White	_____	_____	_____	_____	_____
Totals	_____	_____	_____	$\chi^2 =$ _____	

1. How many degrees of freedom do you have in the interpretation of the χ^2 value?

_____

2. Using Table 3.2, determine what χ^2 values lie on either side of the χ^2 calculated in Table 3.3.

   Record these values here. _____

3. What probability values are associated with these table values of χ^2?

_____

4. In the space provided, write a brief interpretation of the χ^2 value you have just calculated.

_____

_____

_____

_____

5. Place the χ^2 value for your sample in the proper column (under the appropriate P value) of the table on the chalkboard that is prepared in the style of Table 3.4.

TABLE 3.4. χ^2 Values Obtained by Various Class Members for Individual Samples Removed from a Large Population Consisting of Equal Numbers of Colored and White Beans

Degrees of Freedom	$P = .99$	.95	.80	.50	.20	.05	.01
____	____	____	____	____	____	____	____

Note that this table is in the same form as Table 3.2, but space has been provided in the body of the table for recording the χ^2 values obtained by all members of the class. The actual distribution of the χ^2 values from all of the experiments conducted in the class can be observed in the completed table. Thus, although the population is known to be composed of equal numbers of colored and white beans, random samples drawn from it may vary considerably in how closely they approximate the theoretical 1:1 ratio. Nevertheless, in only about 5% of the samples drawn will the χ^2 values be expected to equal or exceed 3.841.[1]

B.   Application of Chi-Square to Monohybrid Data from Investigation 1

Transfer the F_2 data obtained in Investigation 1 (see Table 1.2) to Table 3.5 Calculate χ^2 for the total of males and females based on the hypothesis that the classical monohybrid F_2 ratio has been obtained.

1. How many degrees of freedom do you have in the interpretation of this χ^2 value?

_____

---

[1]   Included in the *Instructor's Manual* is a set of data from 24 students that demonstrates this principle.

TABLE 3.5. Calculation of χ^2 on Data from Investigation 1

Phenotypes	O	E	O − E	$(O - E)^2$	$(O - E)^2/E$
	___	___	___	___	___
	___	___	___	___	___
Totals	___	___	___	$\chi^2 =$ ___	

2. What χ^2 values from Table 3.2 lie on either side of the χ^2 value calculated in Table 3.5?

_____

3. What probability values are associated with these table values of χ^2?

_____

4. In the space provided, write a brief interpretation of the χ^2 value you have just calculated.

_____

_____

_____

_____

C.   Application of Chi-Square to Dihybrid Data from Investigation 1

Transfer the F_2 data obtained in Investigation 1 (see Table 1.5) to Table 3.6 Calculate χ^2 for the total of males and females based on the hypothesis that the classical dihybrid F_2 ratio has been obtained.

1. How many degrees of freedom do you have in the interpretation of this χ^2 value?

_____

TABLE 3.6. Calculation of χ^2 on Data from Investigation 1

Phenotypes	O	E	O − E	$(O - E)^2$	$(O - E)^2/E$
	___	___	___	___	___
	___	___	___	___	___
	___	___	___	___	___
	___	___	___	___	___
Totals	___	___	___	$\chi^2 =$ ___	

2. What χ^2 values from Table 3.2 lie on either side of the χ^2 value calculated in Table 3.6?

_____

3. What probability values are associated with these table values of χ^2?

_____

4. In the space provided, write a brief interpretation of the χ^2 value you have just calculated.

_____

_____

_____

_____

D.   Application of Chi-Square to Data from Investigation 2

Transfer the data concerning tossing of coins from Table 2.1 to Table 3.7 and calculate χ^2. Next, indicate the number of degrees of freedom and the range of P values associated with the χ^2 you have calculated, and conclude whether or not the data are a satisfactory approximation of the expected ratio. Transfer the data from Table 2.2 to Table 3.8, calculate χ^2, and draw appropriate conclusions.

II. CALCULATION OF CHI-SQUARE USING A CALCULATOR

You will increase both the speed and the accuracy with which you calculate χ^2 if you use an inexpensive hand-held calculator for that purpose. A hand calculator suitable for the purposes of this investigation can be purchased for $10 or less.

TABLE 3.7. Calculation of χ^2 on Data from Investigation 2 (Table 2.1)

Results	O	E	O − E	$(O - E)^2$	$(O - E)^2/E$
Heads	_____	_____	_____	_____	_____
Tails	_____	_____	_____	_____	_____
Totals	24	_____	_____	$\chi^2 =$ _____	

Degrees of freedom: _____

P range: _____

I **accept/reject** the hypothesis that the data approximate the expected ratio of 1:1

TABLE 3.8. Calculation of χ^2 on Data from Investigation 2 (Table 2.2)

Results	O	E	O − E	$(O - E)^2$	$(O - E)^2/E$
Heads on both coins	_____	_____	_____	_____	_____
Heads on one, tails on the other coin	_____	_____	_____	_____	_____
Tails on both coins	_____	_____	_____	_____	_____
Totals	36	_____	_____	$\chi^2 =$ _____	

Degrees of freedom: _____

P range: _____

I **accept/reject** the hypothesis that the data approximate the expected ratio

A.   Practice Problems

Use a calculator to calculate χ^2 in the two examples that follow. The χ^2 values you should obtain are given. If the value you obtain does not agree with the value given, repeat the calculation until agreement is reached. Notice how the χ^2 calculated is interpreted.

1. In a cross involving *Drosophila melanogaster*, an F_2 population included 272 flies with long (normal) wings and 60 flies with dumpy wings. Calculate χ^2 and fill in the blanks below. Do these results approximate a 3:1 ratio?

Phenotype	O	E	O − E	$(O - E)^2$	$(O - E)^2/E$
Normal	_____	_____	_____	_____	_____
Dumpy	_____	_____	_____	_____	_____
Totals	_____	_____		$\chi^2 =$	_____

Instructor's calculation of $\chi^2 =$ 8.4980

   a. In interpreting this χ^2 value, you have _____ degrees of freedom.
   b. In this case, do you **accept/reject** the hypothesis that these data approximate a 3:1 ratio?

   _____

   c. What is the probability that the deviations are due to chance alone?

   _____

2. In a dihybrid **testcross**, the following results were obtained: *A–B–*, 121; *A–bb*, 107; *aaB–*, 115; and *aabb*, 93. Do these data approximate a dihybrid testcross ratio with independent assortment? Complete the following table and answer the related questions.

Phenotype	O	E	O − E	$(O - E)^2$	$(O - E)^2/E$
A–B–	121				
A–bb	107				
aaB–	115				
aabb	93				
Totals				$\chi^2 =$	

Instructor's calculation of $\chi^2 =$ 4.0367

a. In interpreting this χ^2 value, you have _____ degrees of freedom.
b. In this case, do you **accept/reject** the hypothesis that these data approximate a dihybrid test-

cross ratio with independent assortment? _____

c. What is the probability that the deviations are due to chance alone?

_____

B.   Assigned Problems

On the following pages are six χ^2 problems to be solved using a calculator. Your instructor will assign one or more of these problems to be submitted at a designated time.

1. The following are dihybrid F_2 data. Do these data support the hypothesis of independent assortment? Complete the table, calculate χ^2, and answer the questions based on your calculations.

Phenotype	O	E	O − E	$(O - E)^2$	$(O - E)^2/E$
A–B–	311				
A–bb	105				
aaB–	91				
aabb	25				
Totals				$\chi^2 =$	

a. In interpreting this χ^2 value, you have _____ degrees of freedom.
b. In this case, do you **accept/reject** the hypothesis that these data approximate the dihybrid F_2 ratio with independent assortment?

_____

c. What is the probability that the deviations are due to chance alone?

_____

d. Complete the following table for the data given. Determine whether each gene pair is behaving according to Mendel's first law, giving a 3:1 ratio. Note that because F_2 data are being considered in this problem, one might reasonably expect each individual trait to behave according to Mendel's law of segregation, giving a 3:1 ratio.

Hypothesis	χ^2 value	P value	Accept or reject hypothesis
3 A–: 1 aa			
3 B–: 1 bb			

2. The following are dihybrid testcross data. Do these data approximate the ratio one would expect for independent assortment? Complete the table, calculate the χ^2, and answer the questions related to your calculations.

Phenotype	O	E	O − E	$(O − E)^2$	$(O − E)^2/E$
A–B–	102				
A–bb	61				
aaB–	71				
aabb	97				
Totals				$\chi^2 =$	

a. In interpreting this χ^2 value, you have _____ degrees of freedom.

b. In this case, do you **accept/reject** the hypothesis that these data approximate the dihybrid testcross ratio with independent assortment?

_____

c. What is the probability that the deviations are due to chance alone?

_____

d. Complete the following table for the data given in this problem. Note that the data in this problem are dihybrid testcross data. Now, determine whether each gene pair is behaving individually as you would expect in a monohybrid testcross. Each trait considered individually should be expected to approximate a 1:1 ratio as a consequence of Mendel's law of segregation.

Hypothesis	χ^2 value	P value	Accept or reject hypothesis
1 Aa : 1 aa			
1 Bb : 1 bb			

e. In view of the χ^2 values obtained for each trait individually, how might you account for the dihybrid testcross ratio obtained?

_____

3. Some studies show that the following are the approximate frequencies of the various ABO blood types in the U.S. white population: 41% A, 9% B, 3% AB, and 47% O. Do the data given in the following table represent a satisfactory sample of the U.S. white population? Complete the table, calculate χ^2, and answer the questions based on your calculations. Calculate E values to the nearest one-hundredth (0.01).

Phenotype	O	E	O − E	$(O - E)^2$	$(O - E)^2/E$
A	599				
B	119				
AB	35				
O	686				
Totals				$\chi^2 =$	

a. In interpreting this χ^2 value, you have _____ degrees of freedom.
b. In this case, do you **accept/reject** the hypothesis that the data given represent the distribution of blood types in the U.S. white population?

_____

c. What is the probability that the deviations are due to chance alone?

_____

d. Determine the percentage of the sample that each blood type constitutes and record this information in the following spaces:

A_____ %; B _____%; AB _____%; O _____%.

4. Probability theory and the binomial expansion (see Investigation 2) show that were you to sample families consisting of four children, 1/16 of these families can be expected to consist of 4 boys, 4/16 would consist of 3 boys and 1 girl, 6/16 would consist of 2 boys and 2 girls, 4/16 would consist of 1 boy and 3 girls, and 1/16 would consist of 4 girls. Do the data in the sample given in the next table approximate this expectation? Complete the table, calculate χ^2, and answer the questions based on your calculations.

Family sex ratio	O	E	O − E	$(O - E)^2$	$(O - E)^2/E$
All boys	168				
3B:1G	646				
2B:2G	919				
1B:3G	568				
All girls	131				
Totals				$\chi^2 =$	

a. In interpreting this χ^2 value, you have _____ degrees of freedom.

b. In this case, do you **accept/reject** the hypothesis that these data approximate the ratio given?

_____

c. What is the probability that the deviations are due to chance alone?

_____

d. Determine whether the overall ratio of boys to girls in the data above is consistent with the hypothesis of a 50:50 sex ratio. Remember that each family included in the table consists of four children; for example, 168 families consisted of 4 boys, 646 families consisted of 3 boys and 1 girl, and 919 families consisted of 2 boys and 2 girls. Calculate χ^2 for these data by completing the following table:

Sex	O	E	O − E	$(O - E)^2$	$(O - E)^2/E$
Male	_____	_____	_____	_____	_____
Female	_____	_____	_____	_____	_____
Totals	_____	_____		$\chi^2 =$	_____

e. **Accept/reject** _____ ; $df =$ _____ ; $P =$ _____

f. Calculate the ratio of boys to girls; record here:_____

_____

g. How have biologists explained sex ratio data such as those observed in this problem?

_____

_____

5. If both parents have the genotype *Aa* and if they produce a family of four children, the binomial expansion shows the probabilities of obtaining families of different phenotypic combinations as follows: 81/256 all *A*–; 108/256 3*A*–: 1*aa* ; 54/256 2*A*–: 2*aa*; 12/256 1*A*–: 3*aa*; and 1/256 all *aa*. A sample was obtained of a large number of families consisting of four children in which both parents were known to be *Aa*. Do these data approximate the expected distribution? Complete the following table, calculate χ^2, and answer the related questions.

Family ratios	O	E	O − E	$(O - E)^2$	$(O - E)^2/E$
All *A*–	1447	_____	_____	_____	_____
3*A*–:1*aa*	1978	_____	_____	_____	_____
2A–:2aa	963	_____	_____	_____	_____
1*A*–:3*aa*	206	_____	_____	_____	_____
All aa	14	_____	_____	_____	_____
Totals	_____	_____		$\chi^2 =$	_____

a. In interpreting this χ^2 value, you have _____ degrees of freedom.

b. In this case, do you **accept/reject** the hypothesis that these data approximate the ratio sug-

   gested above? _____

c. What is the probability that the deviations are due to chance alone?

   _____

d. Note that if the mating $Aa \times Aa$ is made, the theoretically expected ratio among the offspring
   is $3A\!\!-\!\!:1aa$. How many families represented in the table have four children in the expected 3:1

   ratio? _____

   What percentage of the total is this? _____

e. What is the theoretically expected percentage of the families that should have four children in

   a perfect 3:1 ratio? _____

f. How do you account for this low frequency of the expected results?

   _____

6. Binomial expansion tells us that in families of five children you may expect 1/32 of the families
   to consist entirely of boys, 5/32 to consist of 4 boys and 1 girl, 10/32 to consist of 3 boys and
   2 girls, 10/32 to consist of 2 boys and 3 girls, 5/32 to consist of 1 boy and 4 girls, and 1/32 to consist entirely of girls. Do the data on such families of five children given in the following table
   approximate the distribution suggested? Complete the table, calculate χ^2, and answer the related
   questions.

Family sex ratio	O	E	O − E	$(O - E)^2$	$(O - E)^2/E$
All boys	63				
4B:1G	284				
3B:2G	520				
2B:3G	488				
1B:4G	239				
All girls	38				
Totals				$\chi^2 =$	

a. In interpreting this χ^2 value, you have _____ degrees of freedom.

b. In this case, do you **accept/reject** the hypothesis that these data approximate the ratio given

   above? _____

c. What is the probability that the deviations are due to chance alone?

   _____

d. Determine whether the overall ratio of boys to girls in the data is consistent with the hypothesis of a 50:50 sex ratio. Remember that each family included in the table consists of five children; for example, 63 families consist of 5 boys, 284 families consist of 4 boys and 1 girl, and 520 families consist of 3 boys and 2 girls. Calculate χ^2 for these data by completing the following table:

Sex	O	E	O − E	$(O - E)^2$	$(O - E)^2/E$
Male	_____	_____	_____	_____	_____
Female	_____	_____	_____	_____	_____
Totals	_____	_____		$\chi^2 =$	_____

e. **Accept/reject** _____ $df =$ _____; $P =$ _____

f. Calculate the ratio of boy to girls; record here: _____

C. Some Cautions when Using Chi-Square

Some authorities recommend that a certain adjustment be made if the sample in any class is relatively small. For your purposes, however, the calculated value for χ^2 without this correction is satisfactory. Statisticians also suggest that calculating χ^2 is inappropriate when the sample size in any class is less than 5. *With certain exceptions, χ^2 calculations must be based on numerical frequencies and not on percentages or ratios.*

REFERENCES

DYTHAM, C. 2003. *Choosing and using statistics: a biologist's guide*, 2nd ed. Malden, MA: Blackwell Publishing Co.

GILBERT, N. 1989. *Biometrical interpretation: making sense of statistics in biology*, 2nd ed. New York: Oxford University Press.

GLANTZ, S.A. 2001. *Primer of biostatistics*, 5th ed. New York: McGraw-Hill Publishing Co.

KLUG, W.S., and M.R. CUMMINGS. 2006. *Concepts of genetics*, 8th ed. Upper Saddle River, NJ: Prentice Hall.

LEBLANC, D.C. 2003. *Statistics: concepts and applications for science*. Boston, MA: Jones & Bartlett Publishers.

MOSS, R. 1996. A multimedia computer program to teach statistics in genetics. Walking through the chi-square test. *Journal of College Science Teaching* 25(4): 270–273.

PAGANO, M., and K. GAUVREAU. 2000. *Principles of biostatistics*, 2nd ed. Pacific Grove, CA: Brooks/Cole Publishing.

ROHLF, F.J., and R. R. SOKAL 1994. *Statistical tables*, 3rd ed. New York: W. H. Freeman and Co.

ROSNER, B. 2000. *Fundamentals of biostatistics*, 5th ed. Pacific Grove, CA: Brooks/Cole Publishing.

SALKIND, N.J. 2003. *Statistics for people who (think they) hate statistics*. London, UK: SAGE Publications.

SOKAL, R.R. 2005. *Biometry*, 4th ed. New York: W.H.Freeman and Co.

WARDLAW, A.C. 1999. *Practical statistics for experimental biologists*, 2nd ed. New York: John Wiley & Sons.

ZAR, J.H. 1998. *Biostatistical analysis*, 4th ed. Upper Saddle River, NJ: Prentice Hall.

Note. All references cited in Investigation 2 are also pertinent to this investigation.

INVESTIGATION 4

Cell Reproduction: Mitosis

Unfortunately for Mendel, the basic cytology of cell and organismic reproduction was unknown at the time when he was doing his fundamental research on the genetics of the garden pea. Between the publication of Mendel's paper in 1866 and its rediscovery by Correns, DeVries, and von Tschermak in 1900, however, much chromosome cytology came to be understood. In particular, by 1900, the processes of both mitosis and meiosis had been well described. Cytologists learned that during somatic cell division the mitotic process preserves a constant chromosome number, whereas meiosis occurs during gamete formation in animals or spore formation in plants and reduces the chromosome number from diploid to haploid. Because parallels between gene and chromosome transmission could be noted, Mendel's laws were much more meaningful to biologists in 1900 than in 1866.

For example, it was realized shortly after 1900 that the transmission of genes from parents to offspring could be understood satisfactorily only in terms of the way chromosomes behaved in the meiotic process. Then, if the genes were located in the chromosomes, a chromosome theory of inheritance could be postulated. In turn, Mendel's laws of inheritance could be understood and accepted in terms of chromosome behavior determining gene behavior.

Students of genetics, then, do well to begin considering the science of genetics with a study of the cytological basis for gene behavior. Cell reproduction and the mitotic process are a logical starting point.

The ability to reproduce is one of the fundamental properties of life. In this laboratory investigation you will study reproduction at the cellular level. Typically, this process of cellular reproduction involves the division of the nucleus (**mitosis**) to form two identical daughter nuclei, followed by a division of the cytoplasm (**cytokinesis**), which results in the two nuclei being separated into different cells. Mitosis and cytokinesis together constitute cell division, that is, cell reproduction.

The process of mitosis is basically the same in all organisms, both plant and animal. Although most of your observations will be of plants, the generalizations made are applicable to both plants and animals. Where differences exist in the process of cell division in plants and animals, they involve not mitosis per se but the process of **spindle formation** (with centrioles in animals or without centrioles in higher plants) and the mode of cytokinesis (**furrowing** in animals, **cell plate formation** in plants).

OBJECTIVES OF THE INVESTIGATION

Upon completion of this investigation, you should be able to

1. **outline** the procedure for preparing an acetocarmine squash of a root tip to demonstrate the process of mitosis, and

2. **compare** and **contrast** the mitotic process in plant and animal cells.

Materials needed for this investigation:

For each student:

microscope

microscope slides

cover glasses

prepared slide of onion root tip mitosis[1]

prepared slide of human metaphase chromosomes[1]

For each pair of students:

acetocarmine stain in dropping bottle (Dissolve 1.0 g carmine in 100 ml of 45% acetic acid, boil 5 min., cool, decant, and filter.)

1 molar (M) hydrochloric acid (HCl) in dropping bottle

watch glass

dissecting needles

scalpel or razor blade

forceps

alcohol lamp or slide warmer

glass rod for crushing root tips

For the class in general:

supply of onion bulbs[2]

beakers in which to let onions sprout

colchicine solution

absolute ethyl alcohol

70% ethyl alcohol

glacial acetic acid

chloroform

demonstration slide or slides of mitosis in whitefish blastula

I. MITOSIS IN ONION ROOT TIP SQUASHES

A. Onion bulbs will sprout roots if they are placed in water for several days in the manner illustrated in Figure 4.1. The bulbs should be placed in water about 4 days before the laboratory work is to be done. However, individual batches of onion bulbs will respond quite differently to conditions suitable for germination. Many onions obtained commercially have been treated to prevent sprouting. Consequently, they produce very few roots. (A better sprouting of commercial onions can be obtained by carefully slicing a thin layer of tissue from the bottom of the onion bulb.) Onion sets, or locally grown onions, will sprout more effectively.

   Young, actively growing roots that are from 2.5 to 5 cm in length should be used for the study of mitosis. The cells near the tip of the root are actively dividing, thus causing growth.

B. Clip the terminal 1 cm of the root tip from a growing onion bulb during the laboratory period and use it immediately, or use a root tip harvested by the instructor and preserved in Carnoy's

---

[1] Slides are available from Carolina Biological Supply Co., 2700 York Road, Burlington, NC 27215.

[2] With $2n = 12$ and relatively large chromosomes, the broad bean, *Vicia faba*, may be substitued for onion in this investigation. Germinate seeds in layers of moistened paper towels placed in plastic trays. Keep toweling moistened until root tips are harvested.

FIGURE 4.1. Onion bulb supported by tripod of toothpicks in beaker of water. Roots can be removed and actively growing root tips squashed and examined for cells in mitosis.

solution (6 parts absolute ethyl alcohol : 3 parts chloroform : 1 part glacial acetic acid) for 24 hours and then stored in 70% ethyl alcohol until the time of use.

C. To examine the mitotic process in the cells of the onion root tip, you must soften the root so that the cells can be separated and flattened, thus making it possible to see the chromosomes, nuclei, spindles, and other cell parts.

1. Place several drops of 1 M HCl in a watch glass. Be careful not to get the acid on your skin or clothing.
2. Into this acid place the terminal 3 or 4 mm of the 1-cm-long onion root.
3. In a short time (a few minutes) the root tip will feel soft when touched with a dissecting needle.
4. Now, using forceps or a needle, pick up the softened root tip and transfer it to a drop of acetocarmine stain on a clean slide.
5. Using a razor blade or a sharp scalpel cut off and retain the tipmost 1 mm of the root. Discard the rest of the root. Chop the remaining root tip into many pieces with the razor blade and crush the material with a glass rod. *Note:* Iron in the scalpel or dissecting needle reacts with the acetocarmine stain to give a better staining reaction.
6. Once this procedure is complete, apply a clean cover glass to the slide and heat it gently over an alcohol lamp or slide warmer. *Do not boil*! Then invert the slide on a paper towel and push downward firmly, applying pressure with your thumb over the cover glass. This should flatten the cells and disperse them so they can be observed under the microscope.[3]

---

[3] Investigation 6 gives a procedure for making the slide permanent.

7. Examine under low power (100✕) and then under high power (430✕).
   a. Can you locate the various stages of mitosis—prophase, metaphase, anaphase, and telophase? (See Figure 4.2.)

   _____

   b. Most of the cells are in what stage? _____
   c. Those cells having their nuclei intact may show evidence of clear, vacuole-like bodies

   within the nuclei. What are these vacuole-like bodies? _____

D. The mitotic poison **colchicine** can be used to reveal details of chromosome morphology, permit chromosome counts to be made, or induce polyploidy. Colchicine inhibits the formation of the mitotic spindle and thus stops the mitotic process at metaphase stage. Colchicine is an alkaloid compound derived from the corm and other plant parts of the autumn crocus, *Colchicum autumnale*; it was used medicinally in low concentration to treat gout and in higher concentrations as a chemotherapy agent.

   The instructor will provide you with onion root tips that were allowed to grow for a short period of time in a dilute (0.03% to 0.05%) colchicine solution. Follow the same procedure used previously and prepare a squash of the colchicine-treated root. Examine the slide under low power and then under high power.

   1. Do you observe more cells in mitotic stages than you observed in the untreated root tip?

   _____

   2. If so, explain why. _____

   _____

   3. Can you discern anything of chromosome morphology? _____
      a. For example, can you see that some of the chromosomes are composed of two chromatids?

      _____

      b. Can you locate the centromere on any such chromosome? _____

   4. Can you determine the number of chromosomes in any cells? _____

   5. Do you observe any polyploid cells? _____

   6. What mitotic stages are most prevalent? _____

   7. Are any mitotic stages completely lacking? _____

      Why? _____

II. MITOSIS IN LONGITUDINAL SECTIONS OF ONION ROOT TIPS

After completing the laboratory work using squashes of onion root tips, obtain a commercially prepared slide of an onion root tip from the instructor. These slides contain thin (ca. 7 μm to 12 μm) longitudinal sections of onion root tips.[4] They were prepared by a fairly complex process in which the root tip was killed and fixed, dehydrated in alcohol, embedded in paraffin, sectioned on a device

---

[4] Note that 1 micrometer (μm) equals 0.001 millimeter (mm) and is equivalent to the micron (μ).

FIGURE 4.2. Stages of mitosis in plants cells.

called a microtome, stained, and mounted on a microscope slide in Canada balsam or a similar medium. Such a slide may be used repeatedly if it is not damaged by improper handling.

A. Observe the slide using low power and then high power.

    1. How does this slide differ from the ones you prepared? _____

        _____

    2. What advantages do your slides have over the commercial ones? _____

        _____

    3. What are the advantages of the commercial slides? _____

        _____

    4. Carefully study cells in the various mitotic stages. Be certain that you can recognize these stages immediately on sight (Figure 4.2).

    5. Can you see the spindle in these cells? _____

    6. At late telophase, can you see the cell plate? _____

B. Carefully remove the slide from the microscope, clean immersion oil from the slide if the 100× oil-immersion objective was used, and return it to the instructor.

III. MITOSIS IN ANIMAL CELLS

Obtain a slide of mitosis in whitefish blastula and observe with low- and high-power objectives of your microscope. If time is limited, your instructor may set up demonstration slides for your observation. If this is the case, the microscopes will be on high power. Use the fine-focus adjustment only.

    1. Can you observe the spindle? _____ Asters? _____ Centrioles? _____

    2. Do you see any cells undergoing cytokinesis? _____ If so, be sure to observe how cytokinesis occurs in animal cells.

    3. What is a blastula? _____

    4. Briefly summarize the differences between plant and animal cells with reference to mitosis and

       the cell division process. _____

       _____

       _____

IV. OBSERVATION OF HUMAN LEUKOCYTE CHROMOSOMES

The instructor will provide you with a slide of human leukocyte chromosomes. Examine the slide under low power to locate mitotic stages and then observe under high power.

    1. These preparations were made using colchicine. What is the most common mitotic stage found?

       _____

    Why? _____

2. Locate a metaphase stage and observe under high power. Describe the appearance of the chromosomes and their chromatids.

_____

_____

_____

   a. Can you see chromosomes with two chromatids? _____
       Briefly describe the appearance of these metaphase chromosomes and their chromatids.

       _____

       _____

       _____

   b. Can you locate the centromere on any of the chromosomes?

       _____

   c. Is the centromere position the same on all chromosomes?

       _____

   d. Describe the size, shape, and centromere position of several chromosomes._____

       _____

       _____

   e. Can you determine the chromosome number? _____

REFERENCES

ANGYAL, J. 1980. Mitotic cell division. *Carolina Tips* 43(5):21–23.

———. 1981. *Mitosis and meiosis illustrated.* Burlington, NC: Carolina Biological Supply Co.

FLAGG, R.O., and W.R. WEST. 1969. Plant mitosis and cytokinesis. *Carolina Tips* 32(6):21–23.

JOHN, B., and K.R. LEWIS. 1980. *Somatic cell division.* Carolina Biology Reader Series. Burlington, NC: Carolina Biological Supply Co.

MAZIA, D. 1961. How cells divide. *Scientific American* 205(3):100–113.

———. 1974. The cell cycle. *Scientific American* 230(1):54–64.

McFALLS, F.D., and D.G. KEITH. 1969. Animal mitosis. *Carolina Tips* 32(1):1–3.

NASMYTH, K. 2001. Disseminating the genome: joining, resolving, and separating sister chromatids during mitosis and meiosis. *Ann. Rev. Genet.* 35: 673–745.

OUD, O., and G. RICKARDS. 1999. *Understanding mitosis and meiosis: an interactive education tool* (CD-ROM). New York: Springer-Verlag.

PICKETT-HEAPS, J.D., and J. PICKETT-HEAPS. 2002. *The dynamics and mechanics of mitosis.* Video-cassette (60 minutes). Sunderland, MA: Sinauer Associates.

RIEDER, C.L. 1999. *Mitosis and meiosis.* Methods in Cell Biology, vol. 61. New York: Academic Press.

SINGH, R.J., 2003. *Plant cytogenetics.* 2nd ed. Boca Raton, FL: CRC Press.

SUMNER, A.T. 2003. *Chromosomes: organization and function.* Malden, MA: Blackwell Publishing.

TAYLOR, J.H. 1958. The duplication of chromosomes. *Scientific American* 198(6):36–42.

ZIMMERMAN, A.M., and A. FORER, eds. 1981. *Mitosis.* New York: Academic Press.

INVESTIGATION 5

Meiosis in Animals: Oogenesis and Spermatogenesis

In Investigation 4 you studied the process of cell reproduction: a mitotic division of the nucleus followed by cytokinesis. Mitosis produces two nuclei that are genetically and cytologically identical to each other and to the nucleus from which they came. Each mitotic division is preceded by replication of the DNA, and of the chromosomes of which the DNA is a part, so that the daughter nuclei receive identical sets of chromosomes. As a consequence of this very regular process, the body (somatic) cells of an organism all bear the same number and kinds of chromosomes and genes in their nuclei.

In sexually reproducing organisms, a special division process called **meiosis** occurs. Unlike mitosis, meiosis results in the **reduction** of chromosome number; each daughter nucleus has **half the number** of chromosomes found in the somatic cells. Each **haploid** (*n*) nucleus produced through meiosis contains one chromosome of each pair found in the **diploid** (*2n*) cells of the same organism. The details of the meiotic process are described in most general biology and genetics textbooks.

Meiosis takes place at different points in the life cycles of different organisms. In animals, meiosis usually occurs when sex cells (**gametes**) are formed. In plants, meiosis usually occurs when haploid cells called **spores** are formed. (These spores subsequently divide by mitosis to form gametes.) In both plants and animals the diploid condition is restored when gametes unite; for example, in humans the sperm contains 23 chromosomes, the egg contains 23 chromosomes, and the **zygote** resulting from fertilization contains 23 pairs or 46 chromosomes.

OBJECTIVES OF THE INVESTIGATION

Upon completion of this investigation, you should be able to

1. **list** the unique features of meiosis that set it apart from mitosis,

2. **describe** in chronological order the main events of meiosis in ascaris oogenesis,

3. **identify** the early meiotic events of grasshopper spermatogenesis, and

4. **compare** and **contrast** spermatogenesis and oogenesis to identify the unique features of each process.

Materials needed for each student for this investigation:

> microscope
> set of slides of ascaris oogenesis[1]
> slide of grasshopper spermatogenesis[1]

---

[1]   Slides of ascaris oogenesis and grasshopper spermatogenesis are available from the following sources:
Connecticut Valley Biological Supply Co., 82 Valley Road, P.O. Box 326, Southhampton, MA 01073.
Frey Scientific, P.O. Box 8101, Mansfield, OH 44901-8101.
Nebraska Scientific, 3823 Leavenworth St., Omaha, NE 68105-1180.
Sargent-Welch, P.O. Box 5229, Buffalo Grove, IL 60089-5229.

preserved and prestained grasshopper testis

acetocarmine or aceto-orcein stain in dropping bottle

microscope slides

cover glasses

glass rod for crushing grasshopper testis

alcohol lamp or slide warmer

I. MEIOSIS IN ASCARIS OOGENESIS

In today's investigation you will first study meiosis as it occurs during the course of egg production (**oogenesis**) in the roundworm ascaris (*Parascaris equorum*, also called *Ascaris megalocephala bivalens*). This parasitic worm normally lives in the intestinal tract of the horse.

Ascaris is widely used in the study of meiosis because its low chromosome number simplifies observation of individual chromosomes and their behavior. The species of ascaris you are using has two pairs ($2n = 4$) of chromosomes in its diploid cells. A related form has only one pair of chromosomes (i.e., $2n = 2$).

You will be given either one slide containing several (probably five) different sections of the **uterus** of ascaris or four or five different slides, each containing a single section of ascaris uterus. Each section contains cells in a different stage of maturation. The series presents all of the major events of oogenesis and early embryo development in ascaris. Handle the slides with care; they are expensive.

A. Place the slide on the stage with the label to the left. If you have a slide with all of the stages on it, focus on the top row using **low power** (100X). Note the irregularly shaped "eggs," which are about 70 μm in diameter. Actually, these cells are diploid cells in which meiosis has not yet begun or is only about to begin.
   1. What name should be given to these cells?_____

   Careful inspection of these cells under low power will reveal that they often contain one or two small black spots 3 μm to 6 μm in diameter. A perfect "egg" (**primary oocyte**) in this section of the uterus will, in fact, contain two small masses of chromatin representing the **synapsed (paired) homologous chromosomes**.

   Also note the **spermatozoa**, which appear as very black, somewhat triangular (shaped like a flatiron) objects about 20 μm by 8 μm in size. Many sperm cells will be seen among the oocytes. In other cases, a sperm may appear inside the primary oocyte, because in ascaris the sperm penetrates the diploid cell prior to meiosis of the potential egg. The sperm will seldom appear as large within the oocyte as it does when outside the egg. The shape of the sperm will also be altered. An ascaris sperm lacks a flagellum, but instead it uses amoeboid movement (Figure 5.1a).
   2. Why is it important for spermatozoa to have a means of locomotion? _____

   _____

   Carefully shift to **high power** (430X) and repeat these observations. Both pairs of synapsing chromosomes and the sperm can be seen simultaneously only rarely in an oocyte, because the original oocyte was about 70 μm thick; consequently, seven or more thin (10 μm) sections have been cut from each cell. In many of these sections, one, two, or even all three darkly staining bodies have been excluded.

B. Now focus on the second row using low power. The "eggs" in this row are much more definite in shape and each is surrounded by a shell (egg coat) about 3 μm thick. The center of some cells will contain a small black spot that is part of the chromatin of the sperm. Find an oocyte with two black masses near the shell. Now, shift to high power and locate an oocyte in which each of these masses appears to be made of four divisions (**chromatids**). Such a cell (**primary oocyte**) is in

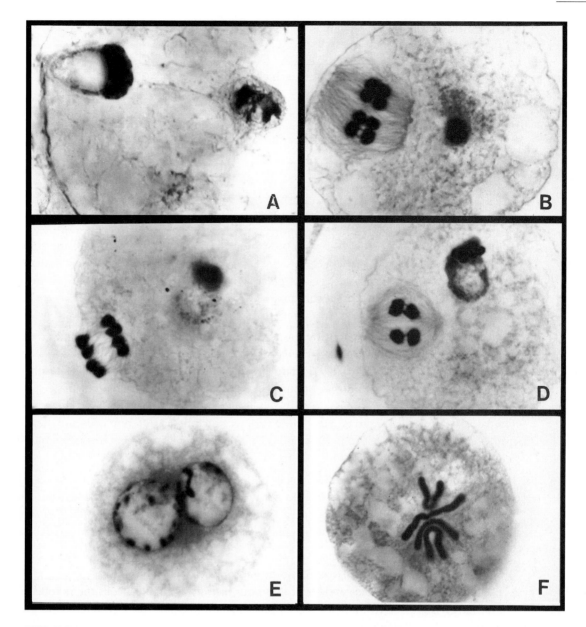

FIGURE 5.1.   Meiosis in ascaris. (a) Sperm entrance; sperm is the triangular body on the left. The diploid (unreduced) nucleus of the primary oocyte is visible to the right. (b) Metaphase of the first meiotic division (metaphase I). Note the two tetrads on the equator of the spindle. (c) Telophase I and the formation of the first polar body. The chromatin from the male gamete is centrally located in the cell. (d) Meiosis II, metaphase in the secondary oocyte. Note the dyads on the equator of the spindle. The remains of the first polar body can be seen on the inside of the shell, opposite the spindle. (e) Male and female pronuclei. (f) Mitosis or first cleavage. Note the diploid number of chromosomes. (Photographs by Carolina Biological Supply Company.)

the tetrad stage. Each chromosome of a pair has become visibly duplicated, producing the tetrad. Some tetrads may be getting ready to divide. A **spindle** can also be seen, with the chromosomes located on it at metaphase I or anaphase I. (See Figures 5.1b and 5.1c.)

C. Focus on row 3. The shell is now about 6 μm in thickness. In either row 2 or, more likely, row 3, you should be able to identify tetrads dividing into two doublet structures (**dyads**). Many oocytes in row 3 will have half of each of the two tetrads apparently attached to the shell as a single black mass. This is the **first polar body**. Such an oocyte is now called a **secondary oocyte**, and each of the double chromatin masses remaining in its cytoplasm is called a **dyad**.

1. How many chromosomes remain in the secondary oocyte?

   _____

2. How many chromatids are present in each chromosome?

   _____

3. How many chromosomes are present in the first polar body?

   _____

4. Are these cells haploid or diploid? _____

   _____

    In other cells of row 3, the dyads can be seen in a new spindle (Figure 5.1d), and some cells may show half of each dyad moving toward the edge of the cytoplasm. The chromatin that moves to the edge of the cell forms the second polar body. The chromatin masses remaining in the cytoplasm are monads or undivided chromosomes. Some such cells, properly called *eggs*, may be seen in rows 3 and 4.

D. Most cells in row 4 show definite nuclei. A perfect specimen of an egg will contain two nuclei: a **female pronucleus**, containing the chromatin of the two remaining chromosomes in a noncondensed or interphase-like stage, and a **male pronucleus**, containing, also in an interphase-like condition, the chromatin brought in by the sperm (Figure 5.1e). In addition, a perfect egg will have its first polar body on the inside of the shell and its second polar body on the edge of the cytoplasm. It may be necessary to view these various structures in different cells and then to construct a composite picture. The perfect egg is called a **mature ovum**. Each was ready to begin cleavage (mitosis) when it was killed and fixed.

E. Now examine the last row of cells. Some cells in this section of the uterus will be zygotes, already beginning cleavage. The two pronuclei do not fuse in interphase. On the contrary, the chromosomes condense, the centrioles elaborate a spindle, and the chromosomes of the two pronuclei become arranged on the spindle in the first mitotic metaphase preparatory to cleavage of the zygote (Figure 5.1f).

   1. Find such a cell at cleavage. How many chromosomes are present? _____

   2. Locate a cell with centrioles and show it to your instructor. In other cells the mitotic division may have progressed to anaphase or telophase. Still others may have progressed to the two-, four-, or eight-cell stage of embryonic development.

F. The instructor will give general directions for preparing drawings. Draw a representative cell from each of the five rows of material. Make drawings of the following five cells: (1) primary oocyte with sperm penetration, (2) primary oocyte containing tetrads, (3) secondary oocyte with first polar body, (4) ovum with male and female pronuclei, and (5) zygote undergoing first cleavage.

    In some cases, it may be necessary to make a composite drawing to show all of the components of the cell. In row 4, for example, it may be necessary to examine several cells to see both male and female pronuclei as well as both polar bodies. The drawing, however, should include all of these components in one cell.

II.   MEIOSIS IN GRASSHOPPER SPERMATOGENESIS

Spermatogenesis is the process of sperm formation in testes of male animals. Spermatogenesis begins when diploid germ cells termed **spermatogonia** differentiate and increase in size to form diploid primary spermatocytes. **Primary spermatocytes** undergo the first meiotic division to form two **secondary spermatocytes**, each of which contains the haploid chromosome number. Each secondary spermatocyte then undergoes the second meiotic division, with the separation of chromatids, to form two **spermatids**. Each spermatid undergoes differentiation and growth of a flagellum to form a functional sperm. Thus, in contrast to oogenesis, spermatogenesis results in the production of four functional gametes from each primary spermatocyte that undergoes meiosis. Grasshopper has often been used in the study of spermatogenesis because of the availability of testes and the beautiful chromosome preparations that can be made, especially of prophase I.

Prophase I of meiosis consists of five substages (Figure 5.2): leptonema (thin thread); zygonema (adjoining threads), during which synapsis occurs; pachynema (thick threads), during which the chromosomes condense longitudinally and **bivalents** or **tetrads** become obvious; diplonema, in which portions of the chromosomes begin to separate and **chiasmata** can be visualized; and diakinesis, which is characterized by further contraction of chromosomes and **terminalization** of pairing. Refer to your genetics textbook for a description of terms with which you are not familiar.

In this section you will make squash preparations of grasshopper testes and study meiosis in males.

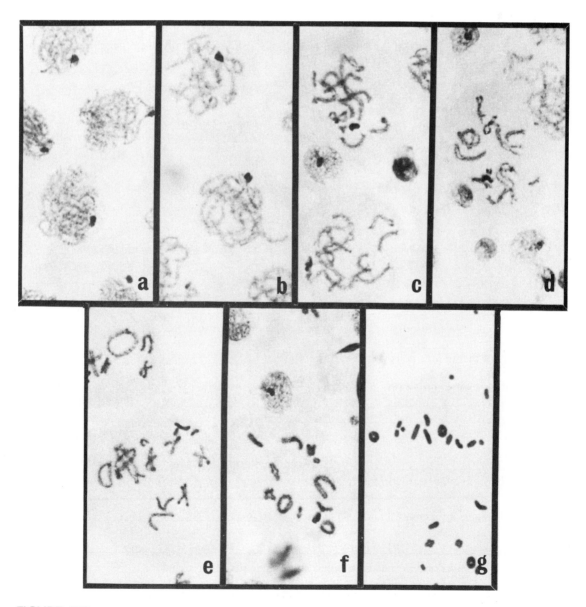

FIGURE 5.2. Meiosis in grasshopper: (a) leptonema, (b) zygonema, (c) pachynema, (d) diplonema, (e) early diakinesis, (f) late diakinesis, (g) metaphase I. (Photographs by R.L. Hammersmith.)

A.   Preparation of Slides

1. Obtain a clean microscope slide and place several drops of acetocarmine stain on it.

2. Your instructor will provide you with a piece of grasshopper testis preserved with Carnoy's solution. Testes may be prestained with Azure A (a stain specific for DNA). A protocol for staining grasshopper testicles is provided in the *Instructor's Manual*. Place the piece of testis in the acetocarmine stain on your slide.

3. Crush the tissue with a glass rod that has been flattened on one end. Break up the tissue into very small pieces.

4. Apply a clean cover glass to the slide and heat it gently over an alcohol lamp or on a slide warmer. **Do not boil**! Then invert the slide on a paper towel and push downward firmly, applying pressure with your thumb over the cover glass. This should flatten the cells and disperse them so you can observe them under the microscope.

5. Examine the slide with your microscope using low power (100×) to locate various stages, and then use high power dry (430×) or oil immersion to study the stages carefully.

6. A technique for making slides permanent is given in Investigation 6.

7. If time does not permit preparation of squashes, your instructor may provide you with a commercially prepared slide of grasshopper testis. These slides generally are of sectioned material rather than squashes.

B.   Observation of Slides

Locate and draw each substage of prophase I, using Figure 5.2 as a guide, and then answer the following questions.

1. What major chromosomal event occurs between leptonema and zygonema? _____

   _____

2. Do any of the chromosomes at zygonema appear to consist of two parallel parts? _____

   How do you account for this appearance? _____

   _____

3. Consult your textbook for a definition of the term **chromomere**. Can you detect chromomeres in

   any of the meiotic cells you are examining? _____

   At what substages of prophase I are chromomeres evident? _____

   _____

4. Do you observe a large, darkly staining structure in the nucleus during leptonema and zygonema?

   _____. This body represents an already highly condensed (heterochromatic) X chromosome. Can you follow the fate of this chromosome through the rest of the

   substages of prophase I and into metaphase I? _____

   _____

5. Briefly list major differences between zygonema and pachynema.

   _____

   _____

6. Locate cells in diplonema. At diplonema, can you observe

    a. the two homologous chromosomes in a pair? _____

    b. individual chromatids in a chromosome? _____

    c. chiasmata? _____

7. Because of the degree of condensation of the chromosomes, diakinesis is an ideal stage at which to determine the chromosome number. Count the chromosomes in a grasshopper cell at diakinesis. Record the number here _____

    Does this represent the diploid number? Justify your answer.

    _____

    _____

    Note that sex in grasshoppers is determined by an XO mechanism in which the female is XX but the male has a single X chromosome. Therefore, the X chromosome that you observe at diakinesis is *not* a tetrad. What is the significance of this information for determining chromosome number in grasshopper males versus females?

    _____

    _____

    _____

8. Observe several cells in metaphase I. Do you notice a chromosome in an unusual position with respect to the other chromosomes in the cell?_____. What chromosome might this be?

    _____

9. Can you find cells in other stages of meiosis or sperm differentiation? _____

    If so, briefly describe their appearance and state what stages you think they might be. _____

    _____

    _____

    _____

III. SUMMARY OF MEIOSIS

Meiosis includes two cycles of division. A total of four haploid nuclei is characteristically produced as a result of meiosis. In oogenesis, these include one functional ovum and two or three polar bodies. The first polar body may divide to give two polar bodies (Figure 5.3). If this happens, the products of oogenesis include one functional ovum and three polar bodies that disintegrate and never function. Because of the way cytokinesis occurs after each meiotic division, the bulk of the cytoplasm with its supply of nutrition is retained in the ovum. The polar bodies are almost devoid of cytoplasm.

In spermatogenesis four cells of equal size (**spermatids**) are produced by the two meiotic divisions. Each spermatid matures into a functional spermatozoon (Figure 5.3).

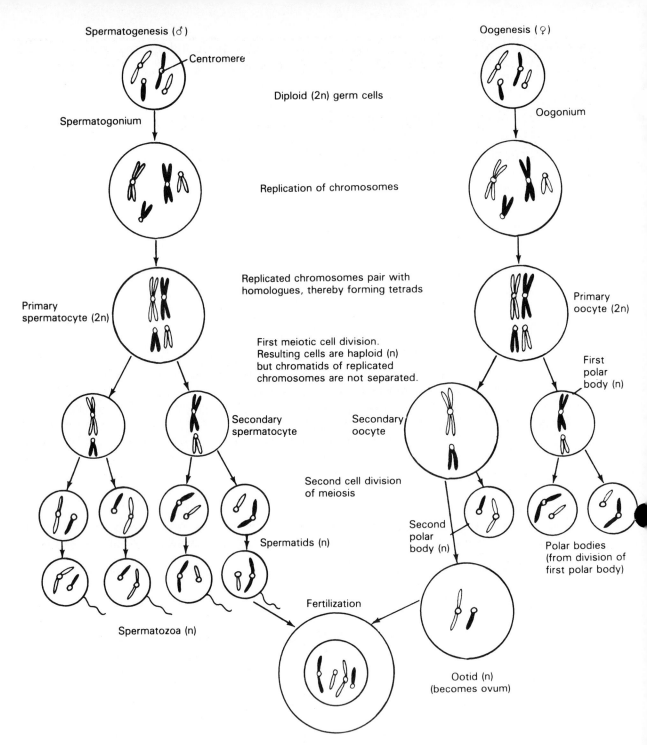

FIGURE 5.3. Comparison of meiosis as it occurs in spermatogenesis and in oogenesis.

REFERENCES

GRIFFITHS, A.J.F., ed. 2004. *An introduction to genetic analysis*, 8th ed. New York: W.H. Freeman and Co.

JOHN, B., and K.R. LEWIS. 1983. *The meiotic mechanism*, 2nd ed. Carolina Biology Reader Series. Burlington, NC: Carolina Biological Supply Co.

JOHN, B. 1990. *Meiosis*. New York: Cambridge University Press.

KLUG, W.S., and M.R. CUMMINGS. 2006. *Concepts of genetics*, 8th ed. Upper Saddle River, NJ: Prentice Hall.

McFALLS, F.D., and D.G. Keith. 1969. Animal meiosis. *Carolina Tips* 32(2):5–7.

MERTENS, T.R., and J.O. WALKER. 1992. A paper-&-pencil strategy for teaching mitosis and meiosis, diagnosing learning problems & predicting examination performance. *The American Biology Teacher* 54(8):470–474.

WAGNER, R.P., M.P. MAGUIRE, and R.L. STALLINGS. 1993. *Chromosomes: A synthesis*. New York: John Wiley & Sons.

INVESTIGATION 6

Meiosis in Angiosperms: Microsporogenesis

As you have already learned, meiosis generally occurs in animals at the time of sex cell (gamete) formation. The gametes unite, form a zygote, and thus initiate the life of a new individual. In plants, however, the process of meiosis usually does not result directly in the production of gametes. Rather, meiosis leads to the formation of **spores**, which then divide mitotically to give rise to **gametophytes** that produce gametes mitotically. The gametes then function in sexual reproduction.

In this investigation you will examine the meiotic process as it occurs in **angiosperms** or flowering plants. Figure 6.1 is a diagram of a typical flower. Actually, meiosis occurs at two locations in such a flower. **Pollen mother cells** located in the **anthers** undergo meiosis and form **microspores**. The haploid nucleus of each microspore subsequently divides mitotically to give rise to a three-nucleate **male gametophyte** (Figure 6.2). Two of the three nuclei are sperm nuclei that function in a double fertilization process.

Meiosis also occurs in a sporecase (**sporangium**) called the **ovule**. This sporecase is located inside the **ovary** of the flower's **pistil** (Figure 6.1). After fertilization occurs, maturation takes place; the ovary develops into a fruit, and the ovules mature into seeds, each of which contains an embryo plant that has developed from a zygote.

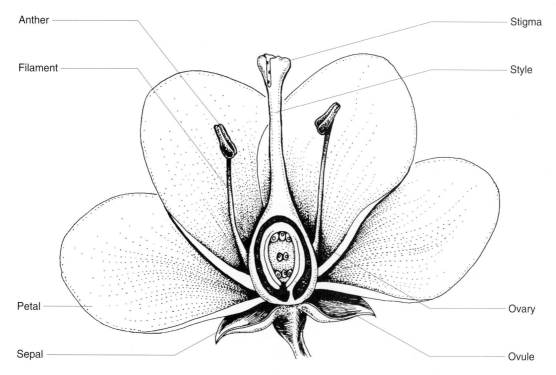

FIGURE 6.1. Parts of a typical flower. Meiosis occurs in the anther and ovule.

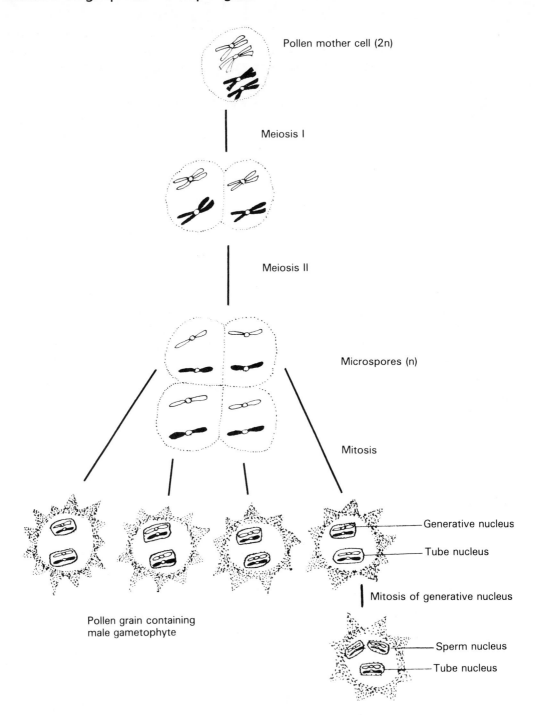

Pollen mother cell (2n)

Meiosis I

Meiosis II

Microspores (n)

Mitosis

Generative nucleus

Tube nucleus

Pollen grain containing
male gametophyte

Mitosis of generative nucleus

Sperm nucleus

Tube nucleus

FIGURE 6.2. In the process of microsporogenesis, one pollen mother cell undergoes two meiotic divisions to form a quartet of microspores. Each microspore matures into a functional pollen grain containing a male gametophyte.

Meiosis in the ovule leads to the formation of a **megaspore**. The megaspore divides mitotically to form a **female gametophyte**, one nucleus of which is an egg. In about 70% of all angiosperms, meiosis in the ovule results in the formation of four megaspores, three of which disintegrate. The one remaining functional megaspore divides mitotically to form two nuclei. Each of these then divides mitotically, resulting in four monoploid (haploid) nuclei. Finally, a third mitotic division produces eight monoploid nuclei, which eventually become partitioned as cells within the female gametophyte.

This female gametophyte is located inside the ovule, and one of its partitioned nuclei serves as the egg nucleus that functions in sexual reproduction. Two of the partitioned nuclei are **polar nuclei** that will be involved in endosperm production. Figure 6.3 shows some of the details of meiosis and female gametophyte formation in a typical flowering plant.

Pollen grains produced in the anther are transferred to the **stigma** of the pistil by insects or wind. There they germinate and begin to grow a pollen tube, which passes down the style and into the ovary of the pistil. The pollen tube carries the sperm nuclei to the ovule. The pollen tube enters the ovule though a pore in the **integuments** called the **micropyle**; once inside the ovule, the pollen tube bursts and frees the sperm nuclei inside the female gametophyte. The pollen tube nucleus then disintegrates.

Double fertilization occurs in angiosperms. One of the sperm nuclei carried into the female gametophyte unites with the egg to form the diploid zygote. This zygote subsequently divides mitotically to give rise to the embryo. The second sperm nucleus unites with the two polar nuclei of the female gametophyte to form the **triploid endosperm**. This cell subsequently divides by mitosis to form additional triploid cells of the nutritive endosperm tissue. The remaining cells (nuclei) of the female gametophyte (the **synergids** and the **antipodals**) generally disintegrate. Thus, the developing seed contains the diploid embryo and the triploid endosperm and is surrounded by the seed coat, which is derived from the integuments of the ovule and is thus maternal in origin.

In this investigation you will study the processes of **microsporogenesis** and male gametophyte formation. **Megasporogenesis** and female gametophyte formation may be studied if your instructor wishes. Ideally, the plant chosen for study should have large chromosomes and a relatively low chromosome number. Different plants have been demonstrated to be especially useful for instructional purposes. Some of these will be mentioned specifically as you read further; additional examples are provided in the references at the end of this investigation.

OBJECTIVES OF THE INVESTIGATION

Upon completion of this investigation, you should be able to

1. **identify** the basic features of meiosis as it occurs in maize microsporogenesis,

2. **outline** the procedure for preparing an acetocarmine squash of anthers from a flower bud to demonstrate the stages of meiosis in microsporogenesis,

3. **identify** the basic features of meiosis as it occurs in a typical angiosperm, and

4. **outline** the alternation of generations as it occurs in the life cycle of a typical angiosperm.

Materials needed for this investigation:

For each student:

compound microscope

stereo dissecting microscope

set of prepared slides of meiosis in maize

microscope slides

cover glasses

watch glass

teasing needles

For each pair of students:

acetocarmine stain in dropping bottle

scalpel or razor blade

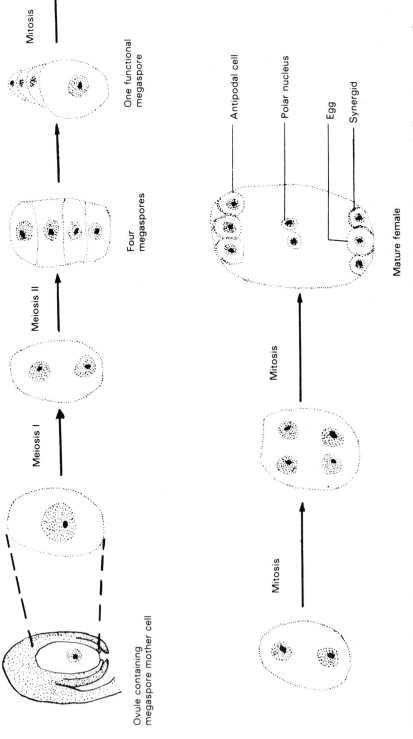

FIGURE 6.3. Megasporogenesis and megagametophyte formation in a typical angiosperm. In about 70% of all angiosperms, only one megaspore is functional, and its nucleus undergoes three mitotic divisions to form an eight-nucleate female gametophyte.

Mitosis

One functional megaspore

Four megaspores

Meiosis II

Meiosis I

Ovule containing megaspore mother cell

Antipodal cell

Polar nucleus

Egg

Synergid

Mature female

Mitosis

Mitosis

forceps

alcohol lamp or slide warmer

For the class in general:

coplin jars

dry ice

euparal mounting medium

supply of immature flower buds

absolute ethyl alcohol

70% ethyl alcohol

glacial acetic acid

chloroform

I.   MICROSPOROGENESIS IN MAIZE

The staminate flowers of Indian corn, or maize (*Zea mays L.*), are located in the tassel at the apex of the plant. During the course of microsporogenesis, meiosis occurs in the anthers of these flowers. Meiosis occurs, however, when the plant is quite young, before the tassel is visible. A complete set of slides of meiosis in maize should include the following 11 stages: leptonema, pachynema, diplonema, diakinesis, metaphase I, anaphase I, telophase I, prophase II, metaphase II, anaphase II, and quartets of microspores in telophase II (Figure 6.4).

If you have access to such prepared slides, handle them with care and examine each using low power and high power of your compound microscope. Your instructor will indicate whether you are to examine the slides using the oil-immersion objective.

As you study the slides of microsporogenesis in maize, answer the following questions.

1. Leptonema, pachynema, diplonema, and diakinesis all occur in cells given what name?

_____

2. a. What are the main differences between cells in leptonema and those in pachynema?

_____

_____

   b.   What accounts for these differences?_____

_____

3. Can you see chromomeres at leptonema?_____

4. During what stage(s) can you see the nucleolus?_____

_____

5. How does diplonema differ from diakinesis?_____

_____

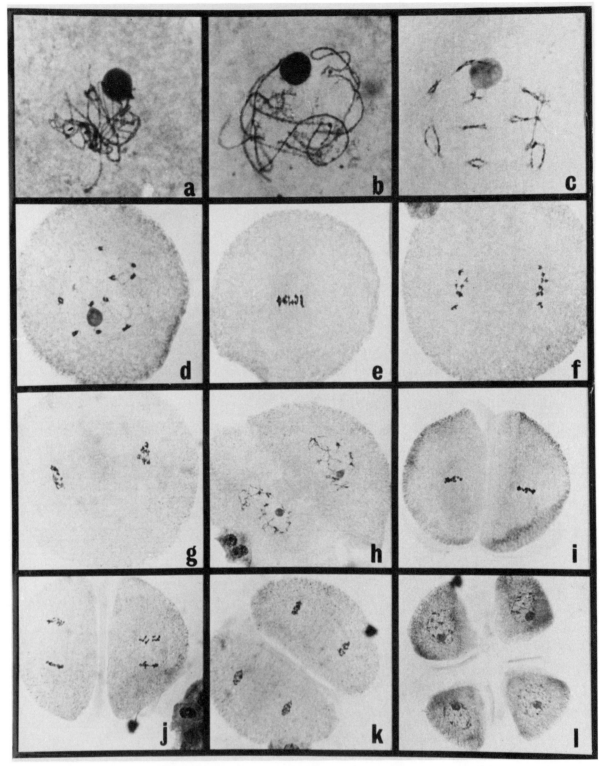

FIGURE 6.4.   Meiosis in maize: (a) leptonema, (b) pachynema, (c) diplonema, (d) diakinesis, (e) metaphase I, (f) anaphase I, (g) telophase I, (h) prophase II, (i) metaphase II, (j) anaphase II, (k) telophase II, (l) microspores. (Photographs by R.L. Hammersmith of slides prepared by R.Y. Hsu, 404 Lakeshore Drive, Lexington, KY 40502.)

6. a. Can you see chiasmata at diplonema and diakinesis?_____

    b. How many chiasmata occur per bivalent (tetrad)?_____

7. a. How many tetrads can you count at diakinesis?_____
    b. How many chromosomes are present in a maize pollen mother cell nucleus at diakinesis?

    _____

    c. Are these cells haploid or diploid?_____

8. How does metaphase I of meiosis differ from metaphase of a mitotic division?_____

    _____

9. a. What is the ploidy of the nuclei formed in telophase I?_____
    b. How many chromosomes will be present in each nucleus at telophase I in maize?

    _____

    c. At prophase II?_____

10. How does prophase II differ from telophase I?_____

    _____

11. a. How can you differentiate between cells in metaphase II and cells in metaphase I?_____

    _____

    b. Anaphase II and anaphase I?_____

    _____

12. At telophase II in the process of microsporogenesis, a quartet of cells is formed.

    a. What name is given to each of these cells?_____

    b. What changes do these cells subsequently undergo?_____

    _____

13. Give the ploidy, and for maize the specific chromosome number, for each of the following cells
    or cell nuclei:

    a. pollen mother cell _____   b. microspore_____

    c. generative nucleus_____   d. tube nucleus_____

    e. sperm nucleus _____

14. Name the plant part in which the cells you have been examining were produced. Be as specific

    as possible._____

    _____

15. Defend the following statement: "In flowering plants such as *Zea mays*, meiosis does not lead directly to the production of gametes (sperm and eggs)."

_____

_____

_____

_____

II. USE OF THE SQUASH TECHNIQUE TO PREPARE SLIDES OF MICROSPOROGENESIS

A. Collecting the Material

Collecting, killing, and fixing developing flowers of various sizes will provide a source of anthers to be used for the study of meiosis by the "squash" technique. Both cultivated and native plant material may be used for this purpose. Among the especially satisfactory or convenient species are maize, rye, *Rhoeo spathacea*, *Tradescantia* spp., wild and cultivated lily species, and *Tragopogon* spp. The usefulness of *Ornithogalum* and chives as angiosperms in which microsporogenesis can be readily studied is also documented in the literature (see References). All of these plants have relatively large chromosomes in comparatively low numbers (Figure 6.5). Another advantage of plants such as maize, *Tradescantia*, *Rhoeo*, *Ornithogalum*, and *Tragopogon* is that one can collect, kill, and fix entire inflorescences (flower clusters), each consisting of numerous lower buds in various stages of meiosis. The plant material must be collected early enough to catch the meiotic divisions while they are occurring. Often the novice will collect too late in the season, after meiosis has been completed and the anthers are filled with mature pollen grains. Experience is the best teacher in this matter.

The collected flower buds must be preserved (fixed) in appropriate chemicals that stop the meiotic process and preserve the chromosomes in their normal form and position. Two of the most widely used fixatives are Carnoy's solution (6 parts absolute ethyl alcohol:3 parts chloroform:1 part glacial acetic acid) and Farmer's fixative (3 parts absolute ethyl alcohol:1 part glacial acetic acid). For best results, these fixatives should always be prepared fresh just before use. Both fixatives produce acid-fixation images; they preserve the chromosomes, nucleoli, and spindle apparatus particularly well.

Materials are placed in vials and completely covered with the appropriate killing-fixing solution, using 50 parts of fixative to 1 volume of material. After material is fixed for 18–24 hours, it may be stored in 70% ethyl alcohol in a refrigerator or freezer. Materials treated in this fashion are effectively preserved for use over a period of years.

B. Preparing a Squash

Remove an entire inflorescence or an individual flower bud from the storage container and place it in a watch glass. Add a few drops of 70% ethyl alcohol to keep the plant material moist. If the flower bud and its anthers are small, you may wish to place the watch glass containing the flower bud on the stage of a stereomicroscope to assist in differentiating the anthers from other flower parts. It is helpful to examine the flower bud against a dark background, because the killed and fixed plant material has been bleached white by the fixative and alcohol. A black card placed on the microscope stage will help you to see better. Continue to keep the inflorescence and the individual flower buds moist in alcohol as you to dissect the anthers from the bud.

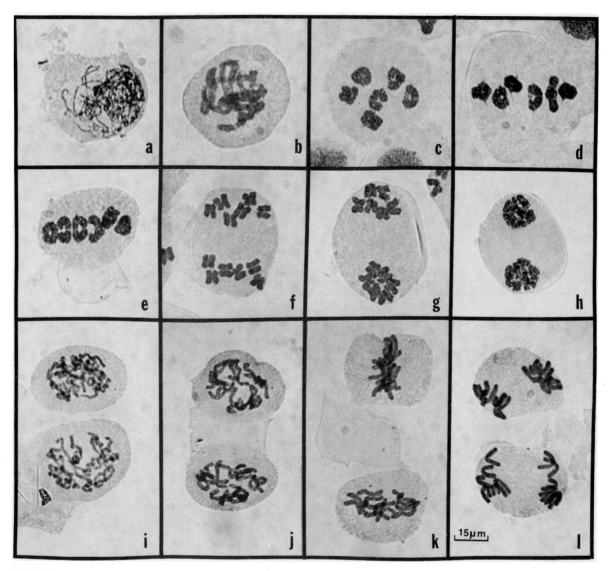

FIGURE 6.5. Microsporogenesis in *Tradescantia*: (a) zygonema, (b) pachynema, (c) diakinesis, (d) metaphase I, (e) metaphase I going into anaphase I, (f) anaphase I, (g) telophase I, (h) late telophase I, (i) early prophase II, (j) late prophase II, (k) metaphase II, (l) anaphase II/telophase II. (Photographs by the authors of acetocarmine squash preparations.)

Now transfer one or two anthers to a clean microscope slide and place them in a drop of aceto-carmine stain. Using a teasing needle, scalpel, or razor blade, macerate the anthers, freeing the microsporocytes from the anther walls. If possible, remove the anther walls from the drop of stain. Apply a cover glass and heat the slide over an alcohol lamp, being careful not to allow the stain to boil. Add stain if necessary to prevent drying.

After heating, cover the slide with paper toweling and press down firmly with your thumb. This pressure will squash or flatten the microsporocytes, making it possible to observe the chromosomes in the various meiotic stages (Figure 6.5). Most of the microsporocytes from any one flower tend to be in about the same stage of meiosis. If anthers from more than one flower are crushed on a slide, different stages are likely to be observed. Examine the slide under the low- and high-power objectives of a compound microscope.

C.   Study of Meiosis in the Course of Microsporogenesis

Study your slide, looking for the various meiotic stages. In microspore mother cells you may expect to see all of the substages of prophase I—leptonema, zygonema, pachynema, diplonema, and diakinesis. Some materials are better for certain stages. For example, maize is especially good for pachytene chromosomes, whereas lily chromosomes effectively reveal the chromomeres that are characteristic of leptonema and zygonema. Metaphase I and anaphase I will also occur in the pollen mother cell. Depending on the species, cytokinesis may or may not occur at telophase I. For example, cytokinesis does occur after telophase I in lily, rye, and *Tradescantia*, but it does not take place in the common garden pepper until after telophase II, when it occurs in two planes, thus producing the four microspores.

TABLE 6.1. Observations of Microsporogenesis in Angiosperms

Record each meiotic stage you have been able to detect in the plants for which you have prepared squashes of anthers. Indicate the name of the plant used in each case.[*]

Meiotic Stage	Name of Plant: _____			Name of Plant: _____		
	Slide 1	Slide 2	Slide 3	Slide 1	Slide 2	Slide 3
Prophase I	_____	_____	_____	_____	_____	_____
Leptonema	_____	_____	_____	_____	_____	_____
Zygonema	_____	_____	_____	_____	_____	_____
Pachynema	_____	_____	_____	_____	_____	_____
Diplonema	_____	_____	_____	_____	_____	_____
Diakinesis	_____	_____	_____	_____	_____	_____
Metaphase I	_____	_____	_____	_____	_____	_____
Anaphase I	_____	_____	_____	_____	_____	_____
Telophase I	_____	_____	_____	_____	_____	_____
Prophase II	_____	_____	_____	_____	_____	_____
Metaphase II	_____	_____	_____	_____	_____	_____
Anaphase II	_____	_____	_____	_____	_____	_____
Telophase II	_____	_____	_____	_____	_____	_____
Mature pollen grains	_____	_____	_____	_____	_____	_____

[*]   You will have to make at least three slides from each plant species investigated if you are to see the various stages of meiosis.

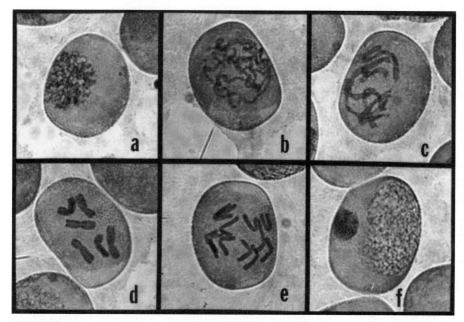

FIGURE 6.6. First mitotic division in a haploid microspore: (a) early prophase, (b) mid-prophase, (c) late prophase (note replicated chromosomes), (d) metaphase, (e) anaphase, (f) two-nucleated pollen grain.

Meiosis II takes place in the nuclei produced as a result of the first meiotic division. Prophase II, metaphase II, anaphase II, and telophase II occur and result in the production of four haploid nuclei, each of which is isolated in a separate microspore by means of cytokinesis. The four microspores separate and become covered by a pollen wall that is characteristic of the species and, thus, constitute four individual pollen grains. In Table 6.1 record your observations about meiotic cells that occur on the slides you have made.

If a slide is made of an anther in which meiosis has already been completed, you might find microspores or mature pollen grains. Although not revealing meiotic stages, such a slide can be useful to demonstrate the first postmeiotic mitosis (Figure 6.6) in which the haploid microspore nucleus divides to form the spherical tube nucleus and the elongate generative nucleus. This mitotic division, as shown in Figure 6.6, can reveal both chromosome number and chromosome morphology.

An occasional anther may produce microsporocytes (Figure 6.7) in which the consequences of chromosome aberrations are evident. Depending on the stages of meiosis present, you might encounter chromosome bridges, acentric fragments, lagging chromosomes, micronuclei, and various combinations of these abnormalities. Be sure to show your instructor any slide you make that reveals such abnormalities as are seen in Figure 6.7. Use your textbook to help you understand the underlying chromosome aberrations that produce these effects.

D. Making a Slide Permanent

To prepare a permanent mount of an acetocarmine squash, use the freezing technique suggested by Conger and Fairchild (1953). Place the slide on a block of dry ice for 60 seconds or more. This procedure freezes the material under the cover glass. While the slide is still on the dry ice, use a razor blade to pry off the cover glass. Most of the cellular material will adhere to the slide; very little will be lost on the cover glass.

Before the stained squash thaws, immerse the entire slide for 5 minutes in absolute ethyl alcohol in a coplin jar. Use two changes of alcohol, with the slide remaining in each change for 2.5 minutes.

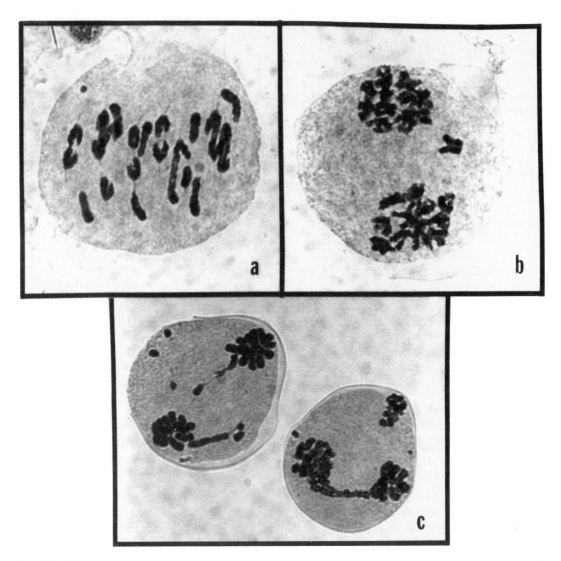

FIGURE 6.7.   Typical chromosome aberrations found in *Tradescantia*: (a) anaphase I where $2n = 24$; (b) late telophase I, with a lagging chromosome; (c) late telophase, showing a chromosome bridge, fragments, and a micronucleus.

Finally, remove the slide from the alcohol and immediately mount the squash in euparal, using a clean cover glass. Take care not to allow the squash to dry before you place the euparal on it. A slide prepared in this way may be used indefinitely. *Note:* This technique may also be used in making squashes of onion root tip, grasshopper testis, and *Drosophila* salivary chromosome preparations permanent.

III.   CONCLUSION

Flowering plants have a distinct alternation of generations, with a monoploid, gamete-producing (gametophyte) generation alternating with an asexual, diploid, spore-producing (sporophyte) generation (Figure 6.8). The sporophyte is the conspicuous flowering plant, while the gametophyte is a microscopic, chlorophyll-free plant that is produced in the anthers (microgametophyte) or ovules (megagametophyte) of the flower.

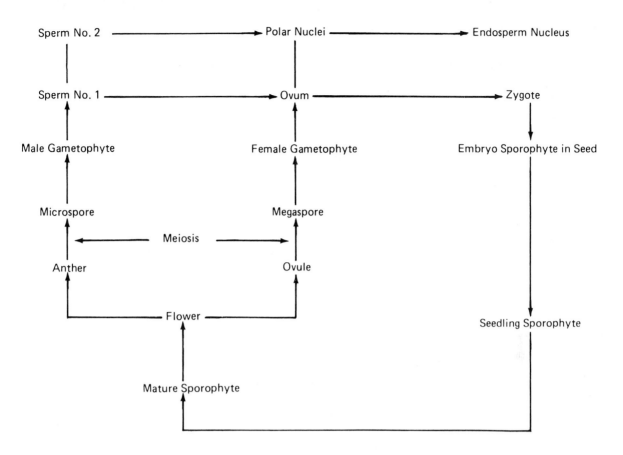

FIGURE 6.8. The life cycle of a typical angiosperm involves an alternation of generations. The sporophyte produces spores by meiosis. The spores divide mitotically to yield gametophytes that contain gametes. Sperm and egg unite to form a zygote, the first cell in the life of the sporophyte generation. Identify the monoploid and diploid phases in the life cycle.

REFERENCES

BEMPONG, M.A. 1973. Meiosis in *Vicia faba*. *Carolina Tips* 36(14):53–54.

CONANT, G.H., and C.W. HAGQUIST. 1944. *Gametophyte development in Lilium michiganese*. Burlington, NC: Carolina Biological Supply Co.

CONGER, A.D., and L.M. FAIRCHILD. 1953. A quick-freeze method for making smear slides permanent. *Stain Technology* 28:281–283.

CRESTI, M., G. CAI, and A. MOSCATELLI, eds. 1999. *Fertilization in higher plants: molecular and cytological aspects*. New York: Springer-Verlag.

FLAGG, R.O., T.E. REGISTER, and W.R. WEST. 1973. Lily life cycle 1. Pollen. *Carolina Tips* 36(10):37–40.

———. 1973. Lily life cycle 2. Embryo sac. *Carolina Tips* 36(11):41–44.

HAMMERSMITH, R.L., and T.R. MERTENS. 1996. *Tradescantia*: a tool for teaching meiosis. *The American Biology Teacher* 59(5):300–304.

JAUHAR, P.P., and W.B. STOREY. 1982. *Ornithogalum virens*: a useful organism for teaching cytology. *Journal of Heredity* 73(3):243–244.

JOHN, B. 1990. *Meiosis*. New York: Cambridge University Press.

KLUG, W.S., and M.R. CUMMINGS. 2006. *Concepts of genetics*, 8th ed. Upper Saddle River, NJ: Prentice Hall.

MAHESWARI, P. 1950. *An introduction to the embryology of angiosperms*. New York: McGraw-Hill Publishing Co.

NASMYTH, K. 2001, Disseminating the genome: joining, resolving, and separating sister chromatids during mitosis and meiosis. *Ann. Rev. Genet.* 35:673–745.

RHOADES, M.M. 1950. Meiosis in maize. *Journal of Heredity* 41(3):58–67.

RIEDER, C.L. 1999. *Mitosis and meiosis* (Methods in Cell Biology, vol. 61). New York: Academic Press.

ROBINSON, A.D. 1982. Teaching meiosis with chives. *Journal of Heredity* 73(5):379–380.

SATTERFIELD, S.K., and T.R. MERTENS. 1972. *Rhoeo spathacea*: a tool for teaching meiosis and mitosis. *Journal of Heredity* 63(6):375–378.

SINGH, R.J. 2003. *Plant cytogenetics*, 2nd ed. Boca Raton, FL: CRC Press.

SUMNER, A.T. 2003. *Chromosomes: organization and function*. Malden, MA: Blackwell Publishing.

INVESTIGATION 7

Polytene Chromosomes from *Drosophila* Salivary Glands

Larval stages of insects of the order Diptera are characterized by having salivary gland cells that contain exceptionally large, multistranded, **polytene** chromosomes. These chromosomes are formed by the process of **endomitosis** in which the chromosomal strands (**chromonemata**) repeatedly replicate but undergo no separation into daughter nuclei. Endomitosis produces exceptionally large, banded, polytene chromosomes (Figure 7.1). The bands that are so evident in these chromosomes are thought to be produced by the lateral association of **chromomeres** of the separate chromosome strands. The bands are rich in DNA, and a combination of appropriate genetic and cytological techniques has demonstrated that certain genes are located in certain bands.

In today's laboratory investigation you will prepare microscope slides of salivary chromosomes using the acetocarmine or aceto-orcein squash technique (similar to that used in Investigations 4 and 6). For this purpose you will use the dipteran *Drosophila virilis*. In other investigations you will use the more common *D. melanogaster*, but because *D. virilis* is larger, it is easier to dissect and remove the salivary glands from larvae of this species.

OBJECTIVES OF THE INVESTIGATION

Upon completion of this investigation, you should be able to

1. **outline** the procedure for preparing an acetocarmine or aceto-orcein squash of *Drosophila* salivary glands for the purpose of demonstrating polytene chromosomes and

2. **recognize** *Drosophila* polytene chromosomes and discuss their significance.

Materials needed for this investigation:

For each student:

binocular stereomicroscope

compound microscope

2 teasing needles

scalpel

microscope slides

cover glasses

For each pair of students:

stock supply of *Drosophila virilis* larvae

dropping bottle of acetocarmine or aceto-orcein stain with dropper

dropping bottle of physiological saline solution with dropper

alcohol lamp

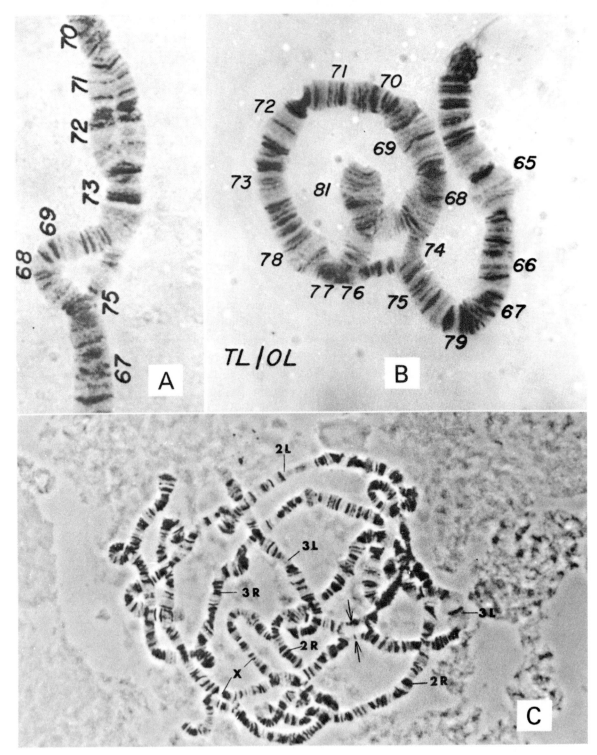

FIGURE 7.1.   Chromosome structural changes as represented in *Drosophila* polytene chromosomes. (a) Heterozygous deficiency in the third chromosome of *D. pseudoobscura*. (b) Heterozygous inversion in the third chromosome of *D. pseudoobscura*. (c) Translocation (see arrows) between the right arm of the second chromosome (2R) and the left arm of the third chromosome (3L) in *D. melanogaster*. (a and b courtesy of C.D. Kastritsis and D.W. Crumpacker, Gene arrangements in the third chromosomes of *Drosophila pseudoobscura*. II. All possible configurations. *Journal of Heredity* 58(1967):112–129. Copyright 1967 by the American Genetic Association. c, courtesy of Burke H. Judd and John Wiley & Sons.)

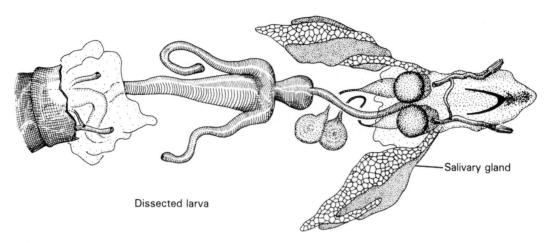

FIGURE 7.2.   Dissection of anterior end of *Drosophila* larva showing the two salivary glands (large cells) with attached fat bodies (stippled).

I.   DISSECTING THE SALIVARY GLANDS FROM THE LARVA

1. Using a teasing needle, remove a large larva from the stock bottle of *D. virilis*. The larger the larva, the easier it will be to dissect. However, be careful not to obtain a larva that has begun to pupate.

2. Place the larva in a drop of physiological saline solution on a clean microscope slide.

3. Place the slide with the larva in the saline solution on a black background on the stage of a stereomicroscope. The white larva and its white or colorless internal organs will contrast well with the black background.

4. Using appropriate magnification (e.g., 15×–25×), dissect the larva. Place one teasing needle in the middle of the larva and the other needle at the anterior end, near the black mouth parts.

5. While holding the larva with the first needle, pull outward with the needle at the anterior end of the larva. This will cause the internal organs of the larva to be pulled out of the body wall.

6. The two salivary glands, which are transparent, are located anteriorly (Figure 7.2). The large cells of the glands give them a characteristic granular appearance at low magnifications. A narrow, opaque ribbon of fat is frequently attached to one edge of the gland. Using teasing needles, remove as much of this fat as possible.

II.   STAINING

1. With a teasing needle, transfer the isolated, fat-free salivary glands to a drop of acetocarmine or aceto-orcein stain on a clean microscope slide. Place a cover glass over the drop of stain.

2. Gently heat the slide over the flame of an alcohol lamp. Take care not to overheat; boiling will destroy the salivary glands.

3. Now cover the slide with a paper towel and push down on the cover glass with the handle of a needle or the eraser end of a pencil. Be careful not to let the cover glass slip while pressing down on it. The pressure you apply will squash the glands, rupture the nuclear membranes, and free the chromosomes so you can view them readily.

4. Observe the slide under the compound microscope, using the low- and then the high-power objective. In a well-prepared slide, the chromosomes will have been expelled from the nucleus so you can see them easily.

   a. Can you detect the banding pattern?_____

   b. Can you see a **nucleolus**?_____

c. If you follow the *Drosophila* chromosomes, can you observe their convergence in a central

location? _____

Such a convergence of the centromeric regions of these chromosomes is termed a **chromocenter**.

d. Can you determine the chromosome number for *D. virilis* from the slide preparations you

have made? _____

Would you expect the salivary gland cells to possess the monoploid or the diploid number of

chromosomes for this species? _____

Why? _____

_____

e. What is meant by **synapsis**? _____

_____

In what kinds of cells do you ordinarily expect synapsis to occur? _____

Why is the question of synapsis raised in connection with the salivary gland chromosomes of

Drosophila? _____

_____

_____

_____

5. Be certain that your instructor sees a slide of salivary gland chromosomes you have prepared.

III. MAKING THE SLIDE PERMANENT

You can make an acetocarmine or aceto-orcein squash of *Drosophila* salivary gland chromosomes permanent by employing the dry ice technique of Conger and Fairchild (see Investigation 6).

IV. STRUCTURAL CHANGES IN CHROMOSOMES

If time permits, and if appropriate *Drosophila* stocks with such chromosome aberrations as deficiencies, inversions, and translocations are available, you may wish to prepare slides and attempt to identify structural changes in the chromosomes (Figure 7.1).

REFERENCES

ASHBURNER, M., K.G. GOLIC, and R.S. HAWLEY. 2005. *Drosophila: a laboratory handbook*, 2nd ed. Cold Spring Harbor, NY: Cold Spring Harbor Laboratory Press.

BEERMAN, W., ed. 1972. *Developmental studies on giant chromosomes*. New York: Springer-Verlag.

DEMEREC, M., ed. 1994. *Biology of drosophila*. Plainview, NY: Cold Spring Harbor Laboratory Press.

DEMEREC, M., and B.P. KAUFMANN. 1986. *Drosophila guide*, 9th ed. Washington, DC: Carnegie Institution of Washington.

LINDSLEY, D.L., G.G. ZIMM, and P.W. HENDRICK. 1987. *The genome of Drosophila melanogaster, Part 3:*

rearrangements. Lawrence, KS: Drosophilia Information Service, No. 65.

LINDSLEY, D.L., and G.G. ZIMM. 1992. *The genome of Drosophila melanogaster*. San Diego, CA: Academic Press.

SORSA, V.V. 1988. *Polytene chromosomes in genetic research*. New York: John Wiley & Sons.

———. 1988. *Chromosome maps of Drosophila*, Vols. 1 and 2. Boca Raton, FL: CRC Press.

ZHIMULEI, I.F. 1998. *Polytene chromosomes, heterochromatin, and position effect variegation*. San Diego, CA: Academic Press.

INVESTIGATION 8

Sex Chromosomes and Gene Transmission

As cytology came to be better understood in the late nineteenth and early twentieth centuries, the role that chromosomes play in sex determination became more apparent. The presence of a particular nuclear structure (the X body) was detected in half the sperms of certain insects as early as 1891. Later (1902), C.E. McClung studied grasshopper species and demonstrated a difference between the chromosome constitutions of males and females: Males always have one less chromosome than females. Grasshopper sperm cells were shown to be of two types, those having a darkly staining (heterochromatic) chromosome (the X body) and those not having such a chromosome. This heterochromatic chromosome came to be known as the X (sex-determining) chromosome (see Investigation 5 for photographs of grasshopper spermatogenesis).

Finally, it became clear that female grasshoppers have two X chromosomes per somatic cell (XX), whereas males have only one (XO). The sex of the offspring is determined by the kind of sperm that fertilizes the ovum; X-bearing sperm cells produce female offspring, and sperm lacking an X chromosome result in male progeny.

Subsequent research revealed that this XO system of sex determination is, in fact, less common than the XY method of sex determination. The latter process involves two heteromorphic, not completely homologous, sex chromosomes. Half the sperm cells produced bear the X chromosome and are female determining, and half bear the Y chromosome and are male determining. Both *Drosophila* and humans have the XY method of sex determination. Research, however, has shown some fundamental differences in the basic genetics of sex determination in these two species despite their similar chromosomal basis for sex differences. In 1916, C.B. Bridges reported that sex in *Drosophila* is determined by a balance between female-determining genes located on the X chromosomes and male-determining genes located on the various autosomes. In humans, and in mammals generally, the male-determining genes are located on the Y chromosome.

In 1910, Thomas Hunt Morgan studied the white eye mutant gene in *Drosophila* in the first investigation to provide extensive experimental evidence for X-linked inheritance. Prior to that time, relationships between sex and gene transmission (e.g., the relationship noted since biblical times for hemophilia) had been observed, but no specific mechanism explaining the relationship had been set forth. Classical hemophilia, of course, is now understood to be due to an X-linked recessive gene. X-linked genes exhibit a crisscross pattern of inheritance in which an allele is transmitted by a male to all of his daughters and from them to approximately one-half of his grandsons.

OBJECTIVES OF THE INVESTIGATION

Upon completion of this investigation, you should be able to

1. **diagram** how an X-linked gene is transmitted from parents to F_1 and F_2 generations in an experimental mating of *Drosophila*,

2. **describe** the effect of mandatory nondisjunction associated with attached-X chromosomes on sex determination and gene transmission in *Drosophila*, and

3. **trace** the transmission of an X-linked gene in a human pedigree.

Materials needed for this investigation:

For each student:

materials for handling *Drosophila* (see Investigation 1)

For the class in general:

wild-type stock of *Drosophila*

stock of *Drosophila* carrying an appropriate X-linked mutant

"attached-X" stock of *Drosophila*

I.   SEX LINKAGE IN *DROSOPHILA*

For this part of the investigation you may use any one of many X-linked mutants of *Drosophila*. The following discussion is based on the assumption that you will use the X-linked white eye (*w*) mutant. Different members of the class will be asked to do reciprocal crosses so that the results can be compared. Review the procedures for making *Drosophila* matings (see Investigation 1).

A.   Cross

Some of you will mate a white-eyed virgin female fly ($X^w X^w$) to a wild-type male ($X^+ Y$).[1] Produce F_1 and F_2 generations. Record data in Table 8.1.

B.   Reciprocal Cross

Others will mate a wild-type virgin female ($X^+ X^+$). to a white-eyed male ($X^w Y$). Produce F_1 and F_2 generations. Record data in Table 8.1. Then compare the F_1 and F_2 phenotypes and genotypes of the reciprocal crosses by completing Table 8.2.

Write a brief summary paragraph describing how the trait you have been studying (white eyes) would be inherited in reciprocal crosses if it were controlled *not* by a gene that is X-linked but rather by one that is autosomal.

II.   MATING INVOLVING AN ATTACHED-X FEMALE *DROSOPHILA*

In the mitotic and meiotic processes, occasional errors will occur when chromosomes fail to separate at anaphase and the two members of a pair (in meiosis I) or the daughter chromatids of a single chromosome (in meiosis II or in mitosis) go to the same spindle pole. Such failure of chromosomes to

---

[1]   Gene symbols shown as superscipts to a capital X represent X-linked genes.

TABLE 8.1. Record of *Drosophila* Experiment

Record data of cross involving a sex-linked gene.

Experiment number _____ **Name** _____

1. Cross _____ female $\times$ _____ male

2. Date P_1s mated: _____

3. Date P_1s removed: _____

4. Date F_1s first appeared: _____

5. Phenotype of F_1 males: _____

6. Phenotype of F_1 females: _____

7. Date F_1 male and female placed in fresh bottle: _____

8. Date F_1 flies removed: _____

9. Date F_2 progeny appeared: _____

10. Record F_2 data in the following table:

	Males		Females		
	Phenotype	Number	Phenotype	Number	Total
a.	_____	_____	_____	_____	_____
b.	_____	_____	_____	_____	_____
c.	_____	_____	_____	_____	_____
d.	_____	_____	_____	_____	_____
	Totals	_____		_____	_____

separate is known as **nondisjunction**. Special *Drosophila* stocks exist in which compulsory nondisjunction of the X chromosomes occurs because the two X chromosomes of the *Drosophila* female are physically joined together at their centromere ends. Such an "attached-X female" may have a Y chromosome as well and thus have the chromosomal condition $\widehat{XX}Y$. Such a female is viable and fertile. When meiosis occurs, she produces two kinds of eggs, those bearing two (attached) X chromosomes and those bearing Y chromosomes.

1. When such a female is mated to a normal XY male, what chromosomal constitutions can be expected among the resulting zygotes?

_____

_____

TABLE 8.2. A Comparison of the Results of Reciprocal Crosses Involving an X-Linked Gene

	Cross (A)		Reciprocal Cross (B)	
	Phenotype	Genotype	Phenotype	Genotype
P_1 female	_____	_____	_____	_____
P_1 male	_____	_____	_____	_____
F_1 female	_____	_____	_____	_____
F_1 male	_____	_____	_____	_____
F_2 females	_____	_____	_____	_____
F_2 males	_____	_____	_____	_____

2. Would you expect all of these zygotes to be viable and to result in the production of adult flies?

_____ The zygotes having what chromosomal constitution(s) might be suspected of not being viable? _____

3. What is unusual about the way in which the male offspring of such a cross (i.e., $\widehat{XX}Y \times XY$) come into being? _____

_____

Now, mate a virgin wild-type female fly having attached-X chromosomes to a male that expresses an X-linked mutant gene such as that for white eyes (w). Produce F_1 and F_2 generations. Record data relative to this experiment in Table 8.3.

4. Give the genotypes of the P_1, F_1, and F_2 flies from the cross you performed. Show X-linked genes as superscripts on the X chromosomes.

P_1 female _____

P_1 male _____

F_1 female _____

F_1 male _____

F_2 female _____

F_2 male _____

5. On the basis of the results you obtained in this experiment, can you now say with certainty which zygotes resulting from a cross of $\widehat{XX}Y$ to XY are viable and which are inviable? Explain. _____

TABLE 8.3. Record of *Drosophila* Experiment

Record data resulting from the cross of a wild-type attached-X female to a male carrying an X-linked mutant.

Experiment number _____ **Name** _____

1. Cross _____ female × _____ male

2. Date P_1s mated: _____

3. Date P_1s removed: _____

4. Date F_1s first appeared: _____

5. Phenotype of F_1 males: _____

6. Phenotype of F_1 females: _____

7. Date F_1 male and female placed in fresh bottle: _____

8. Date F_1 flies removed: _____

9. Date F_2 progeny appeared: _____

10. Record F_2 data in the following table:

Males		Females		
Phenotype	Number	Phenotype	Number	Total
a. _____	_____	_____	_____	_____
b. _____	_____	_____	_____	_____
c. _____	_____	_____	_____	_____
d. _____	_____	_____	_____	_____
Totals	_____		_____	_____

6. What would happen in the next generation were you to allow the F_2 males and females from this cross to mate *inter se* (among themselves)? What genotypes and phenotypes might be expected in the next generation? _____

_____

_____

7. On the basis of what you learned in this experiment, comment on the role of the Y chromosome in sex determination in *Drosophila*.

_____

_____

_____

_____

III. X-LINKAGE IN HUMANS

A.   Hemophilia

Figure 8.1 shows the results of a study of four generations of a family from Muncie, Indiana, among whom **hemophilia** has occurred. In this diagrammatic representation, circles represent females and squares represent males. Individuals having hemophilia are shown in black, and unaffected individuals are shown in outline only. Note that all affected individuals are males and that these males are the products of a marriage between unaffected parents. Also note that no affected individuals occur in the fourth generation.

The inheritance pattern exhibited in this pedigree is the classical pattern characteristic of hemophilia A, which is caused by an X-linked recessive gene. In the spaces that follow, give the **genotypes** of the various individuals in the pedigree shown in Figure 8.1. Show the alleles as superscripts to the X chromosomes, with $+$ or H representing the normal allele and h the allele for hemophilia.

I _____   _____
     1   2

II _____ _____ _____ _____ _____
      1   2   3   4   5

     _____ _____ _____
       6   7   8

III _____ _____ _____ _____ _____
       1   2   3   4   5

     _____ _____ _____ _____
       6   7   8   9

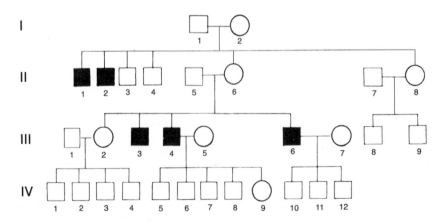

FIGURE 8.1.   Human pedigree showing inheritance of hemophilia, an X-linked gene. Circles represent females; squares represent males. Black squares represent hemophiliacs.

IV _____   _____   _____   _____   _____
       1   2   3   4   5

  _____   _____   _____   _____   _____
       6   7   8   9   10

  _____   _____
      11   12

Indicate the meanings of the gene symbols and genotypes used in answering the previous question._____

_____

_____

If you were a genetic counselor, how would you counsel individual IV-9? _____

_____

_____

_____

What is the probability that she carries the allele for hemophilia? _____

_____

What is the probability that she will have sons who are hemophiliacs? _____

_____

What is the probability that she will have daughters who are hemophiliacs? _____

_____

What is the probability that she will have daughters who are heterozygous carriers of the allele for hemophilia?

_____

_____

B.   Other X-Linked Genes

McKusick (OMIM, 2005) noted that over 907 genes or traits in humans appear to be produced by X-linked loci, and in September, 2005, 548 gene loci had been definitely assigned to the X chromosome.[2] In addition to classical hemophilia A, some of the well-known X-linked recessive genes include

---

[2] Online Mendelian Inheritance in Man (OMIM) is a valuable source for information on human genetics (see References for Web-site address).

those producing various types of color blindness, glucose-6-phosphate dehydrogenase deficiency, Lesch-Nyhan syndrome, Duchenne muscular dystrophy, and ichthyosis. A few X-linked dominant genes have also been identified. An X-linked dominant gene appears to be the cause of vitamin D-resistant rickets (hypophosphatemia). The production of a certain blood antigen, the X_g antigen, is also due to a dominant X-linked gene (symbolized as X_g^a). The recessive allele X_g fails to result in the production of the antigen.

Because they have no alleles on the Y chromosome, the pattern of inheritance for X-linked genes is somewhat different from that for autosomal genes. Consider the following questions pertaining to the inheritance of X-linked genes.

1. Would you expect X-linked recessive alleles such as the ones for hemophilia and color blindness to be more frequently expressed in males or in females? _____

   Why? _____

   _____

2. Would you expect an X-linked dominant allele such as X_g^a to be more commonly expressed in males or in females? _____

   Why? _____

   _____

3. If a woman having normal color vision marries a color-blind man and they have one child, a color-blind son, from whom did the son inherit color blindness: his mother, his father, or both parents? Justify your answer. _____

   _____

   _____

   _____

   _____

4. Historically, the X chromosome had more genes assigned to it than its physical size would seem to warrant relative to the other chromosomes.[3] Can you suggest another explanation as to why this was the case? _____

   _____

   _____

   _____

   _____

---

[3] Today, because of molecular genetic techniques, this difference in the number of assigned genes between the X chromosome and the autosomes is less apparent.

REFERENCES

GELEHRTER, T.D., F.S. COLLINS, T.F. GELEHRTER, and D. GINSBURG. 1998. *Principles of medical genetics,* 2nd ed. New York: Lippincott, Williams & Wilkins.

CUMMINGS, M.R. 2006. *Human heredity: principles and issues.* 7th ed. Belmont, CA: Thomson Brooks/Cole.

LAWN, R.M., and G.A. VEHAR. 1986. The molecular genetics of hemophilia. *Scientific American* 254(3):48–54.

MCCONKEY, E.H. 1993. *Human genetics. The molecular revolution.* Boston: Jones and Bartlett Publishers.

MCKUSICK, V.A. 1962. On the X chromosome of man. *Quarterly Review of Biology* 37:69–175.

———. 1965. The royal hemophilia. *Scientific American* 213(2):88–95.

———. 1998. *Mendelian inheritance in man. Catalogs of human genes and genetic disorders,* 12th ed. Baltimore MD: Johns Hopkins University Press.

MORGAN, T.H. 1910. Sex limited inheritance in *Drosophila. Science* 32:120–122.

NOVARTIS FOUNDATION SYMPOSIUM. 2002. *The genetics and biology of sex determination—No. 244* (CIBA Foundation Symposia Series). New York: John Wiley & Sons.

NUSSBAUM, R.L., R.R. McINNES, and H.F. WILLARD. 2004. *Thompson & Thompson genetics in medicine, revised reprint,* 6th ed. Philadelphia PA: W.B. Saunders Company.

Online Mendelian Inheritance in Man, OMIM™. McKusick-Nathans Institute for Genetic Medicine, Johns Hopkins University (Baltimore, MD) and National Center for Biotechnology Information, National Library of Medicine (Bethesda, MD). 2000. World Wide Web: http://www.ncbi.nlm.nih.gov/omim/

SOLARI, A.J. 1994. *Sex chromosomes and sex determination in vertebrates.* Boca Raton, FL: CRC Press, Inc.

VOGEL, F., and A.G. MOTULSKY. 1997. *Human genetics. Problems and approaches,* 3rd ed. New York: Springer-Verlag.

WACHTEL, S.S., ed. 1994. *Molecular genetics of sex determination.* New York: Academic Press.

INVESTIGATION 9

The Sex Check: A Study of Sex Chromatin in Human Cells

Barr and Bertram (1949) noticed a small, darkly staining body, adjacent either to the nucleolus or to the nuclear membrane, in most of the nuclei of neurons of female cats. This body was rarely observed in similar cells from male cats. Further studies indicated that this sexual dimorphism is present in cells of most tissues from representative marsupials, artiodactylans, chiropterans, carnivores, and primates. In some orders, such as the rodents and other lagomorphs, this sex differential is readily observed in only a few tissues.

This chromatin body, which is characteristic of female cells but is absent from male cells, is now commonly known as *sex chromatin* or as the *Barr body*, after its discoverer. In cells other than nerve cells, this body (about 1 μm in diameter) typically assumes a planoconvex shape, lying adjacent to the inner surface of the nuclear membrane. Cytochemically, sex chromatin is Feulgen positive and stains deeply with methyl green pyronin, indicating the presence of DNA.

Considerable evidence has been amassed indicating that sex chromatin is an inactivated X chromosome. The normal human male, who has only one X chromosome, does not exhibit sex chromatin (Figure 9.1a). The normal human female, who has two X chromosomes, has one sex chromatin body per nucleus (Figure 9.1b), whereas a female trisomic for the X chromosome exhibits two sex chromatin bodies (Figure 9.1c). Rare females who have four X chromosomes have three Barr bodies per nucleus.

The sex chromatin body visible in the interphase nuclei of normal females is due to the inactivation of a major portion of one of the two X chromosomes. This inactivation involves the heterochromatization of that particular X chromosome. Using radioactive labeling techniques, Morishima et al. (1962) found that

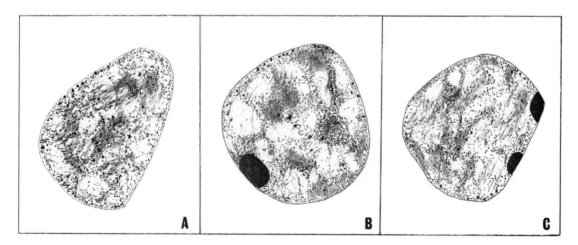

FIGURE 9.1. Nuclei of cells illustrating sex chromatin or Barr bodies. (a) Nucleus of an XY male cell having no sex chromatin. (b) Female nucleus (XX) having one sex chromatin body. (c) Trisomic-X female nucleus having two sex chromatin bodies.

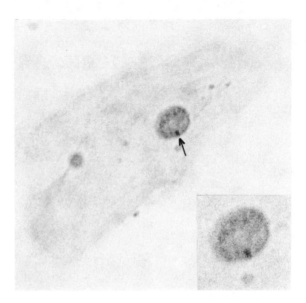

FIGURE 9.2. Human squamous epithelial cell, the nucleus of which exhibits a single sex chromatin body (arrow). Inset shows an enlargement of the same nucleus.

the inactivated X chromosome, as with other **heterochromatin**, replicates later than the other chromosomes in the nucleus. In addition, the late-replicating X is characteristically found at the edge of the nucleus, where the sex chromatin body is most frequently observed. In summary, all cells contain one X chromosome with genetically active chromatin known as **euchromatin**. All additional X chromosomes become heterochromatic sex chromatin bodies.

The procedure for demonstrating sex chromatin in squamous epithelial cells of human oral smears (Figure 9.2) was developed by Ludwig and Klinger (1958). This and related procedures were adopted by the International Olympic Committee for verification of female athletes but not without controversy (Elsas et al., 2000; Puffer, 2002). A modification of the Ludwig and Klinger procedure (Reynolds, 1960) is used in this investigation.

OBJECTIVES OF THE INVESTIGATION

Upon completion of this investigation, you should be able to

1. **outline** the procedure for preparing a microscope slide of human squamous epithelial cells to demonstrate sex chromatin and
2. **describe** the relationship between the number of X chromosomes present in the nucleus of a human cell and the number of sex chromatin bodies evident in that nucleus.

Materials needed for this investigation:

For each student:

microscope equipped with oil-immersion objective

tongue depressor

For each pair of students:

1% aqueous solution of thionin stain

ethyl alcohol in the following concentrations: 50%, 70%, 95%, 100%

xylene (xylol)

5 or 6 normal (N) hydrochloric acid (HCl) in dropping bottle

balsam mounting medium

microscope slides

cover glasses

coplin jars or dishes

forceps for handling clean slides

I.   PROCEDURE

A.   Obtaining the Smear

You may experience two difficulties in preparing a suitable smear from the oral mucosa. If the slide is not scrupulously clean, the epithelial cells will be lost in the course of the staining procedure. Consequently, the slides should be washed in a detergent, rinsed thoroughly in distilled water, and given a final rinse in 70% ethyl alcohol. The slide may then be flamed dry. A second problem, which is more difficult to overcome, is that of contamination of the epithelial cells by oral bacteria. Bacteria are especially troublesome, because they stain readily with thionin and may be confused with sex chromatin if, by chance, they are superimposed on the nucleus of an epithelial cell. To remedy this problem, cell donors should vigorously rinse their mouths with tap water to remove loose epithelial cells and bacteria. This rinsing should be repeated two or three times for best results.

The smear is prepared by gently scraping the lining of the cheek with a tongue depressor and smearing the material thus obtained on a clean slide. The slide should be air dried for 30 seconds and then processed according to the following schedule.

B.   Staining and Mounting

Place each slide in 95% ethyl alcohol for 2 minutes, then in 70% ethyl alcohol for 2 minutes, then in 50% ethyl alcohol for 2 minutes, and finally in distilled water for 2 minutes.

Upon removing the slide from the distilled water, place several drops of 5 or 6 N HCl on the cells for approximately 5 seconds. This is a very critical part of the procedure preparatory to staining. The acid hydrolyzes RNA in the nucleus of the cell and thus eliminates RNA-rich bodies, which also stain with thionin. Excessive hydrolysis may disrupt the DNA as well as the RNA, whereas insufficient hydrolysis may leave such RNA-rich bodies as the nucleoli intact. Thionin is a differential stain for the nucleic acids, staining DNA a dark blue-black and RNA violet. When studying objects as small as the nucleoli and sex chromatin bodies in these cells, however, it is desirable to avoid having to distinguish between the two on the basis of color differences. Proper hydrolysis will eliminate RNA-rich bodies and prevent their being confused with the DNA-rich sex chromatin body.

Next, transfer the slide to distilled water for 10–15 seconds to remove the hydrochloric acid. Then, stain the slide in 1% aqueous thionin for 10–15 minutes, after which time again rinse the slide in distilled water and then dehydrate it according to the following schedule: 50% ethyl alcohol for 30 seconds, 70% ethyl alcohol for 30 seconds, 95% ethyl alcohol for 30 seconds, and absolute ethyl alcohol for 30 seconds.

Now, transfer the slide to xylene for clearing for 1 minute. Upon removing the slide from the xylene, immediately place a drop of balsam mounting medium over the cells and apply a clean cover glass. Take care in placing the cover glass to avoid trapping bubbles in the mounting medium.

Examination of smears prepared in this way should reveal cells with virtually colorless cytoplasm and a pale blue nucleus. The sex chromatin bodies, most frequently located adjacent to the nuclear membrane (Figure 9.2), should be stained a dark blue-black. Locate cells to be examined with the low-power objective; then check for sex chromatin using the oil-immersion objective. Carefully avoid getting balsam on the microscope objective.

II. SEX CHROMATIN AND NUCLEAR SEX

Chromatin bodies resembling sex chromatin are occasionally observed in tissues taken from males. The statistical incidence of what appears to be sex chromatin in male cells normally ranges from about 2% in oral smears to as high as 20% in liver preparations. In fact, for an accurate determination of nuclear sex, a statistically significant number of cells must be examined and the percentage of cells having sex chromatin computed. For oral smears made from females, most investigators find that between 50% and 65% of the healthy cells demonstrate sex chromatin. Only normal-looking nuclei that stain evenly should be scored. Nuclei that are folded, have granulated nucleoplasm, or demonstrate any sign of degeneration cannot be included in a critical study.

1. Examine 50 squamous epithelial cells from a human female. How many exhibit sex chromatin?

   _____ What percentage of these 50 cells reveal sex chromatin? _____

2. What is the incidence of sex chromatin among 50 oral mucosal cells obtained from a human male?

   _____

3. Did you find any cells of the human female that appear to have more than one sex chromatin

   body? _____

4. For most mammals studied, and for humans in particular, evidence has accumulated to show that the Y chromosome is male determining and that, with rare exceptions, whenever it is present the phenotype of the individual will be male. Female-determining genes are apparently located on both the X chromosome and the autosomes. The normal human female has a somatic chromosome complement of 46, consisting of 44 autosomes and two X chromosomes; the normal male has 44 autosomes plus an X and a Y chromosome. On the basis of this information, complete Table 9.1, indicating the probable sex chromosome complement and the total number of chromosomes anticipated in the somatic cells of the individual (assuming the presence of the normal 44 autosomes).

TABLE 9.1. Sex Chromatin and Its Relationship to Sex Chromosome Complement and Chromosome Number

Number of Sex Chromatin Bodies Observed in Somatic Cells	Phenotypic Sex of Individual from Whom Cell Was Obtained	Probable Sex Chromosome Complement	Total Number of Chromosomes in Somatic Cells
0	Male (normal)	_____	_____
0	Female (abnormal)	_____	_____
1	Male (abnormal)	_____	_____
1	Female (normal)	_____	_____
2	Male (abnormal)	_____	_____
2	Female (abnormal)	_____	_____
3	Male (abnormal)	_____	_____
3	Female (abnormal)	_____	_____

5. Refer to your textbook in seeking answers to the following questions.

   a. Which of the individuals depicted in Table 9.1 has classical Turner syndrome? _____

   b. Which has classical (the most common form of) Klinefelter syndrome? _____

   c. Which individuals might be regarded as metafemales? _____

REFERENCES

BARR, M.L., and E.G. BERTRAM. 1949. A morphological distinction between neurones of the male and the female, and the behavior of the nucleolar satellite during accelerated nucleoprotein synthesis. *Nature* 163:676–677.

CUMMINGS, M.R. 2006. *Human heredity: principles and issues*, 7th ed. Belmont, CA: Thomson Brooks/Cole.

ELSAS, L.J., et al. 2000. Gender verification of female athletes. *Genetics in medicine* 2(4):249–254.

LUCCHESI, J.C., W.G. KELLY, and B. PANNING. 2005. Chromatin remodeling in dosage compensation. *Annu. Rev. Genet.* 39:615–651.

LUDWIG, K.S., and H.P. KLINGER. 1958. Eine einfache und sichere Farbemethode fur das sex chromatin Korperchen und dessen feinere Struktur. *Geburtshilfe u. Frauenkunde* 4:555–558.

LYON, M.F. 1972. X-chromosome inactivation and developmental patterns in mammals. *Biological Review of the Cambridge Philosophical Society* 47:1–35.

MITTWOCH, U. 1963. Sex differences in cells. *Scientific American* 209(1):54–62.

———. 1973. *Genetics of sex differentiation.* New York: Academic Press.

MORISHIMA, A., M.M. GRUMBACK, and J.H. TAYLOR. 1962. Asynchronous duplication of human chromosomes and the origin of sex chromatin. *Proceedings of the National Academy of Sciences* (U.S.) 48:756–763.

PLATH, K., S. MLYNARCZYK-EVANS, D.A. NUSINOW, and B. PANNING. 2002. *Xist* RNA and the mechanism of X chromosome inactivation. *Annu. Rev. Genet.* 36:233–278.

PUFFER, J.C. 2002. Gender verification of female Olympic athletes. *Medicine and Science in Sports and Exercise* 34(10):1543.

REYNOLDS, W.A. 1960. A comparison of cytological techniques demonstrating sex chromatin and its occurrence in the raccoon (*Procyon lotor*) and in the little brown bat (*Myotis lucifugus*). Master's thesis, State University of Iowa.

REYNOLDS, W.A., and T.R. MERTENS. 1964. The sex check. *The American Biology Teacher* 26(6): 411–415.

SOLARI, A.J. 1994. *Sex chromosomes and sex determination in vertebrates.* Boca Raton, FL: CRC Press, Inc.

STRACHAN, T., and A.P. READ. 2003. *Human molecular genetics*, 3rd ed. New York: Garland Science/ Taylor & Francis Group.

SUMNER, A.T. 2003. *Chromosomes: organization and function.* Malden, MA: Blackwell Publishing.

INVESTIGATION 10

Human Chromosomes

Chromosomes were observed microscopically by several biologists in the latter part of the nineteenth century. Edward Strasburger described mitosis in living plant cells in 1875. In 1879, Walther Flemming followed the mitotic process in fixed and stained tissues from amphibian larvae. He coined the terms now used to describe the process of mitosis. W. Waldeyer named the major structures visible in the metaphase stage, *chromosomes*. Theodor Boveri and Walter S. Sutton in 1902 associated the genetic material with chromosomes. As cytological techniques improved, chromosomes were observed and studied in many animal and plant cells.

Studies of human chromosomes, however, progressed slowly, because materials were not readily available and techniques that had been applied to plant and animal cells could not be used in humans. These difficulties were resolved when human cells were grown in culture outside the body. Dividing cells in culture could be treated with the alkaloid colchicine, which frees the chromosomes from the spindle apparatus of the dividing cell. Cells in culture can then be exposed to a hypotonic solution that causes the cell to swell, thereby separating the chromosomes so that they can be observed individually and counted.

When these techniques were applied by J.H. Tijo and A. Levan (in 1956) to cultured cells of human lung embryonic tissue, the human chromosome number was established at $2n = 46$. This observation was soon confirmed in England by C.E. Ford, who studied gonadal tissue. Other studies that confirmed the human chromosome number as 46 were based on cell cultures from bone marrow and skin biopsies. The significance of the number of chromosomes present in humans became apparent in 1959 when J. Lejeune and his coworkers attributed the disorder Down syndrome to an abnormal chromosome number. Since that time, many major physical and intellectual consequences have been linked to human chromosomal aberrations.

Investigations of human chromosomes routinely employ blood leukocytes, which are convenient to obtain, culture, and induce to undergo mitosis. When appropriately prepared, human chromosomes can be shown to vary in length (the largest is about 10 μm long; the smallest is about 2 μm long) and by the position of the **centromere (primary constriction)**. A **karyotype** (karyogram[1]) is an assemblage of metaphase chromosomes of an individual arranged in pairs according to the order of descending length and position of the centromere.

OBJECTIVES OF THE INVESTIGATION

Upon completion of this investigation, you should be able to

1. **describe** the morphology of human chromosomes with reference to size, centromere position, and presence or absence of satellites,

---

[1] In late 2005, the term "karyotype" was replaced by the term "karyogram" (ISCN 2005). Students should be aware that in most literature the older term has been used, but they can expect to see "karyogram" used in future publications. In this manual we have elected to continue using karyotype.

2. **describe** representative structural chromosome aberrations using standard numerical and letter symbols, and

3. **prepare** a karyotype of human leukocyte chromosomes and determine number of chromosomes present, sex of the individual, and presence or absence of a numerical or structural chromosome aberration.

I. MORPHOLOGY OF CHROMOSOMES

On the basis of chromosome length and position of the centromere, normal human chromosomes have been arranged in seven groups of autosomes, A to G, and one pair of sex chromosomes, XX or XY. With these criteria, the 23 pairs are classified as follows:

Group	Chromosomes
A	1 to 3
B	4 and 5
C	6 to 12
D	13 to 15
E	16 to 18
F	19 and 20
G	21 and 22
X	XX or XY

Use Figure 10.1 to answer the following questions.

1. A **metacentric** chromosome is one that has a centrally located centromere and chromosome arms with approximately equal length. Which of the human chromosomes are metacentric? Answer by giving individual chromosome numbers. _____

_____

2. **Submetacentric** chromosomes have centromeres located close to the center of the chromosome, but the chromosome arms are distinctly unequal in length. Which human chromosomes fall into this category? _____

_____

3. **Acrocentric** chromosomes have centromeres located much nearer one end than the other, resulting in chromosome arms that are decidedly unequal in length (one of the chromosome arms will be *very* short). Which of the human chromosomes are acrocentric? _____

_____

No human chromosomes have centromeres on the end (**telocentric**), thus having only one arm.

4. Certain human chromosomes also have **secondary constrictions** that usually represent the site of ribosomal RNA genes (nucleolar organizing region; NOR). Morphologically, the secondary constriction separates a portion of the short arm from the main body of the chromosome. These short, distal, chromosomal portions are called **satellites**. (Do not confuse these chromosome satellites and secondary constrictions with satellite DNAs, which are a category of repetitive DNA.) Which of the human chromosomes have satellites? _____

_____

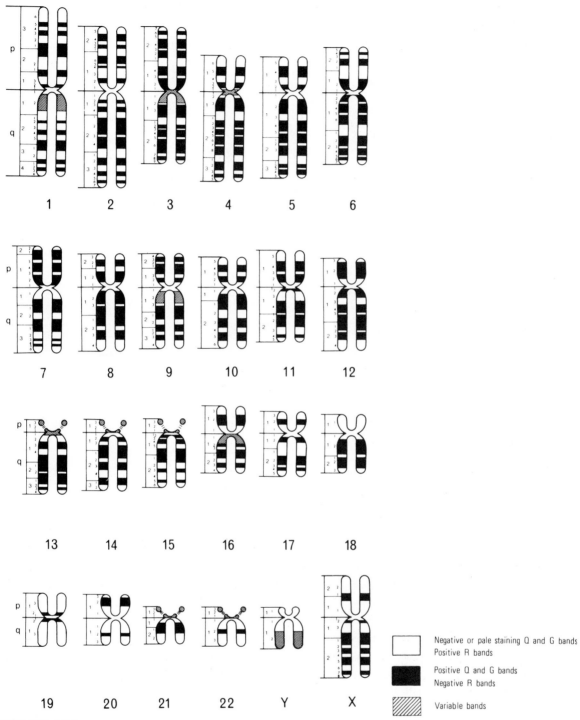

FIGURE 10.1. Chromosome bands as observed with Q-, G-, and R-staining methods. Autosomes are numbered in order from 1 to 22, and Y and X are listed separately. Short arms of chromosomes are designated p and long arms q. Regions are numbered from 1 (next to centromere) to the distal end of each arm, and bands are numbered within the regions from the centromere end to the distal end of each arm. (Paris Conference [1971]: Standardization in human cytogenetics. Birth Defects: Original Article Series, VIII[7], 1972. The National Foundation, New York.)

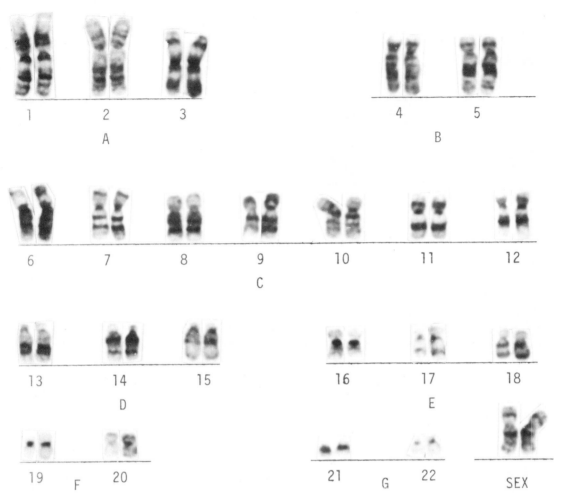

FIGURE 10.2.   Karyotype of human metaphase chromosomes from a leukocyte culture. Chromosomes are arranged in order by length and centromere position into seven groups of autosomes and one pair of sex chromosomes. Banding patterns are used to identify the two chromosomes that are the members of each pair. This karyotype is of a normal 46,XX female. (Courtesy of R.M. Fineman and S.R. Woodward.)

5. Observe the relative length and centromere location of the chromosomes shown in Figure 10.1. Draw brackets around and assign the appropriate letter to each of the eight groups in Figure 10.1. Note that, in terms of size and centromere position, the X chromosome is similar to the chromosomes in the C group of autosomes and the Y chromosome is similar to the G autosomes.

   All human chromosomes have two arms, one on either side of the centromere. For identifying arms of chromosomes with respect to the centromere, p and q are used to designate the short (p) and the long (q) arms, respectively.

6. Using Figure 10.1 as a guide, locate the centromere region of each chromosome in Figure 10.2, place brackets around each arm (centromere to each end) of all chromosomes, and then label each arm with either p or q.

II. BANDED CHROMOSOMES AND STRUCTURAL ABERRATIONS

Staining techniques have been developed to show cross bands on chromosomes. Through these techniques, individual chromosomes within the eight groups can be distinguished with accuracy. In Figure 10.1 three different banding techniques—G, R, and Q—have been superimposed to show the composite of bands. Giemsa blood stain is used to reveal G bands (Figure 10.2). Segments of metaphase chromosomes that are stained to appear darker or lighter than adjacent parts are used to identify particular chromosomes and parts of chromosomes. An alternative method for Giemsa staining reveals opposite banding patterns compared to the standard G banding. These reversed Giemsa bands are called R bands. Fluorescent Q bands are due to staining with quinicrine compounds. Fluorescence can be observed with a specially equipped microscope using ultraviolet light. Chromosome pairs are distinguished by the degree of intensity and other characteristics of quinicrine bands. Usually only one type of banding—for example, G banding—is needed for identification of all metaphase chromosomes in a cell. Two chromosomes that are members of the same pair exhibit similar banding patterns.

The molecular techniques of **FISH** and **chromosome painting** using fluorescently labeled chromosome-specific DNA probes and *in situ* hybridization (see Investigation 11) now augment and complement standard cytogenetic techniques in the detection and analysis of both aberrant karyotypes and complex chromosome rearrangements, some of which are also diagnostic of certain types of cancer (see Speicher et al., 1996). The back cover shows a human leukocyte stained with the FISH (fluorescent *in situ* hybridization) procedure and having a complex chromosome rearrangement involving chromosomes 22 (pink) and either 9 or 12 (blue). See Investigation 11 for a more in-depth discussion of FISH and chromosome painting as well as a more complete description of the cover photograph.

A standard nomenclature, including regions and bands, has been developed for localizing and distinguishing each area of each chromosome. A **region** is an area of a chromosome between two major landmarks such as centromere, conspicuous band, or chromosome end. Regions are numbered in order from the centromere to the end of each arm. **Bands** within each region are numbered starting with the band closest to the centromere and proceeding toward the end of each chromosome arm. In Figure 10.1, the larger numbers at the side of each chromosome refer to the regions of each arm, and the smaller numbers specify bands within each region.

1. In Figure 10.2, locate and number the **regions** for each arm of chromosomes 5, 7, 11, and X. Use

    Figure 10.1 as a guide. Can you distinguish any bands within these chromosomal regions?_____

    _____

2. If so, for which chromosome(s)? _____

If a structural chromosome aberration has occurred in one member of a pair, it may be identified by a mismatch with the homologous banded chromosome. Structural changes within chromosomes or between different chromosomes can therefore be identified with G-banded chromosomes. Structurally altered chromosomes are designated with single-letter or three-letter abbreviations of names for rearrangements, for example, t (**translocation**), del (**deletion**), der (**derivative chromosome**), dup (**duplication**), inv (**inversion**), and ins (**insertion**). Immediately following the type of rearrangement, the number of the chromosome or chromosomes involved is given in parentheses, for example, inv(2), t(2;5). If one of the rearranged chromosomes is a sex chromosome, it is listed first; otherwise, the chromosome with the lower number is specified first, for example, t(X;2) or t(14;21). The location of any given break is specified by the number of the region and band in which the break occurred; for example, when only the region is known, 1p2 (chromosome 1, short arm, region 2), or when a specific band is involved, 1q13 (chromosome 1, long arm, region 1, band 3). For example, a deletion in the long arm of chromosome 2 could be produced by a break at region 3, band 3. This deletion would be symbolized as del(2)(q33). When breaks are in two arms of a single chromosome, the break point in the short arm is specified before the break point in the long arm; for example, inv(3)(p2q3) is a **pericentric** inversion (including the centromere) in chromosome 3 with break points in region 2 of the short arm and region 3 of the long arm. (Inversions within an arm of a chromosome are called **paracentric** inversions).

Abnormalities in structure of a particular chromosome may produce specific phenotypic consequences if the chromosome material becomes unbalanced, resulting in a partial aneuploidy. Unbalanced structural rearrangements occur in approximately 1 in 3300 newborns. A deletion in the short arm of chromosome 5 (5p) is associated with the cri du chat (cat try) syndrome. When the specific region of a deletion or addition of chromosomal material is not specified, a minus (−) or plus (+) can be placed after the symbol for the chromosomal arm. For example, cri du chat syndrome can be symbolized as del (5p−).

Balanced structural chromosome rearrangements occur in approximately 1 in 230 newborns. Many individuals with balanced rearrangements can go unidentified for their entire lives; others receive a diagnosis following infertility, pregnancy loss, or the birth of a child with an unbalanced structural chromosomal abnormality such as cri du chat syndrome of Down syndrome caused by a translocation. An example of a balanced rearrangement is a 45, XX t(14;21) (see Figure 10.3) female who has a normal phenotypic appearance because all genes are functionally diploid. For a reference on frequencies of chromosome aberrations, see Gardner and Sutherland (2003).

Years of study of structural aberrations using refined techniques for identification and analysis of chromosome bands have necessitated a refinement in the nomenclature for specifying certain aberrations (see ISCN, 1995). For example, "a derivative chromosome (der) is a structurally rearranged chromosome generated either by a rearrangement involving two or more chromosomes or by multiple aberrations within a single chromosome" (ISCN, 1995). The term **derivative** (der) is always applied to a chromosome with an intact centromere. Figure 10.3 is a karyotype of a 45,XX female with a balanced robertsonian translocation [t(14;21)] involving chromosomes 14 and 21. The translocation thus produced the derivative chromosome designated as der (14;21) (q10; q10). The complete designation for this individual is 45,XX, der (14;21) (q10; q10).

With appropriate symbols, describe the following chromosome rearrangements.

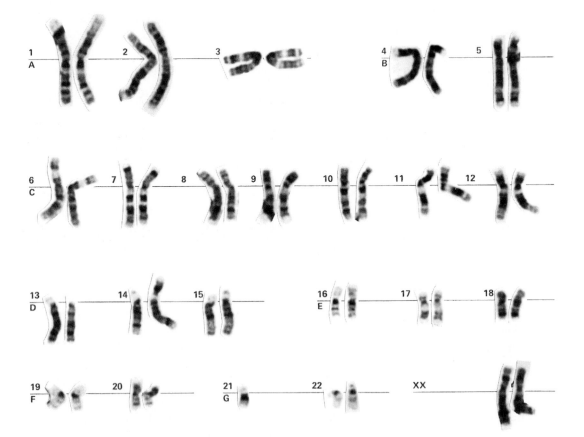

FIGURE 10.3.   Karyotype of a woman carrying the derivative chromosome 14;21 as a result of the translocation of the long arm of chromosome 21 to the short arm of chromosome 14. Although having only 45 chromosomes, this woman is phenotypically normal, because virtually all of chromosome 21 is preserved in the derivative chromosome. She is also at risk for producing a child with Down syndrome. Can you explain why? (Courtesy of D.L. Van Dyke and J. Zabawski, Henry Ford Hospital, Detroit.)

3. A deletion in region 2, band 5 in the long arm of chromosome 4.

_____

4. Paracentric inversion (with two breaks in the same arm) in the long arm of chromosome 6, region 1, with break points in bands 2 and 6.

_____

5. Translocation of the long arm of chromosome 21 to the short arm of chromosome 14 with the retention of the chromosome 14 centromere. Assume a break in the short arm of chromosome 14

at region 1, band 1 and the loss of the entire short arm of 21. _____

The chromosome resulting from this translocation is properly referred to as a _____ chromosome.

6. A pericentric inversion in chromosome 2 with break points in region 1, band 4 of the short arm and region 2, band 3 of the long arm.

_____

7. Studies in recent years have shown that a number of forms of cancer are associated with specific structural chromosome aberrations. For example, chronic myelogenous leukemia (CML) is associated with the so-called Philadelphia chromosome (Ph). The defective chromosomes in CML occur only in white blood cells and involve a reciprocal translocation (exchange of segments) between the long arms of chromosomes 9 and 22 (see Figure 10.4). Careful studies indicate that the

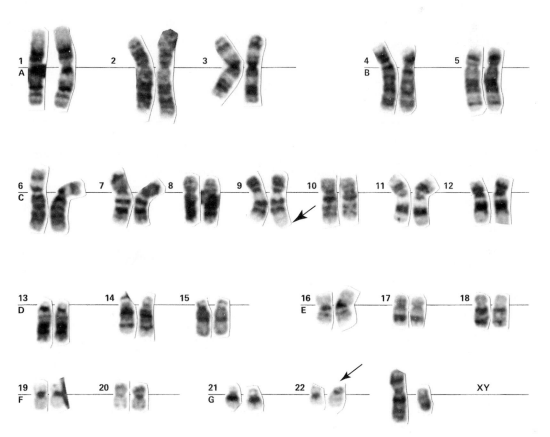

FIGURE 10.4. Karyotype of a leukocyte from a person having chronic myelogenous leukemia associated with a reciprocal translocation between chromosomes 9 and 22 (arrows). See text for further details. (Courtesy of D.L. Van Dyke and J. Zabawski, Henry Ford Hospital, Detroit.)

chromosome breaks occur in the long arms of chromosome 9 (region 3, band 4) and 22 (region 1, band 1). Use proper symbolism to designate the resulting derivative chromosomes.

_____

_____

III. NUMERICAL CHROMOSOME ABERRATIONS

Although the usual number of chromosomes in human somatic cells is 46, additions or losses of whole chromosomes occur in approximately 1 in 300 newborns. Numeric chromosome abnormalities (aneuploidy and polyploidy) are more commonly found in pregnancy losses, due to the often lethal imbalance of chromosome material, and occur in approximately 50% of all first trimester miscarriages. Trisomy 16, which is nonviable, and monosomy X (Turner syndrome) are the most common aneuploidies found in spontaneous abortuses. While most trisomies have been documented in spontaneous absortions, those involving chromosomes 13, 18, 21, X, and Y (and rarely 9 and 22) are able to survive and often have phenotypic consequences. Some aneuploidies are associated with a complex of abnormalities that may include morphological, physiological, and psychological deviations from normal. Such a complex of symptoms is called a **syndrome**. Some of the more common and well-known syndromes associated with chromosomal abnormalities are listed in Table 10.1. Refer to your textbook for a description of the phenotypic effects associated with each of these syndromes.

In attempting to symbolize or describe the chromosome complement of an individual, chromosome number as well as structure must be taken into consideration. Thus, the description of each chromosome complement begins with the total number of chromosomes represented in the sample photograph. Conferees at the 1971 conference in Paris agreed that a plus or minus sign should be placed after the total number but before the appropriate symbol to signify added or missing whole chromosomes. A male having an extra chromosome 13 would be symbolized 47,XY+13 (Figure 10.5).

TABLE 10.1. Variations in Chromosome Number, Barr Bodies, and Corresponding Phenotypes and Syndromes*

Phenotype or Syndrome	Total No. of Chromosomes	Sex Chromosome Complement	No. of Barr Bodies per Nucleus
Normal female	46	XX	1
Normal male	46	XY	0
Male with Klinefelter syndrome	47, 48, 49	XXY, XXXY, XXXXY	1, 2, 3
Female with Turner syndrome	45	X	0
Male with XYY karyotype	47	XYY	0
Female with trisomy X	47	XXX	2
Down syndrome (trisomy 21)	47	XX or XY	1 if female; 0 if male
Edwards syndrome (trisomy 18)	47	XX or XY	1 if female; 0 if male
Patau syndrome (trisomy 13)	47	XX or XY	1 if female; 0 if male

*The reader is referred to general and human genetics textbooks for a full description of the syndromes listed in this table.

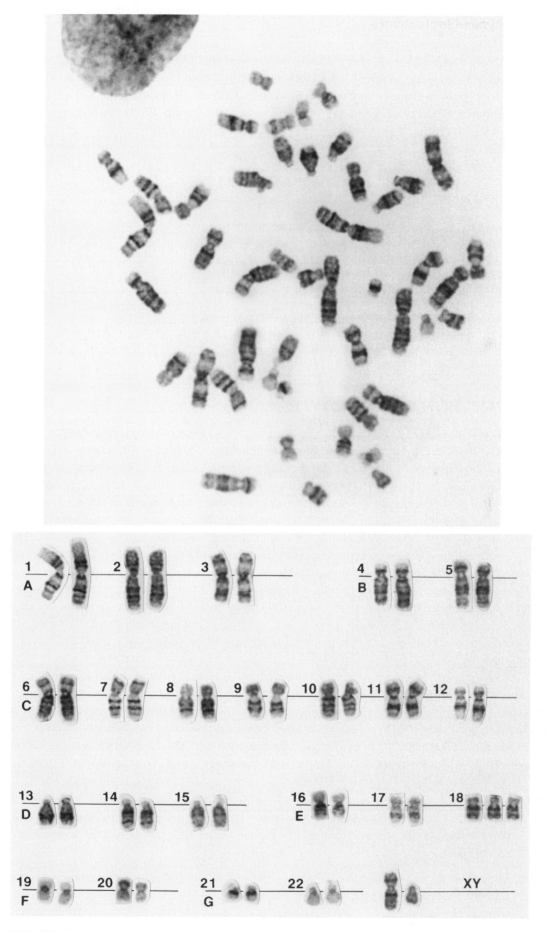

FIGURE 10.5. Chromosome spread (A) and karyotype (B) of a male having trisomy-18, Edwards syndrome (47,XY+18). (Courtesy of A. Wiktor, J. Zabawski, and D.L. Van Dyke, Department of Medical Genetics, Henry Ford Health System, Detroit, MI.)

The conferees agreed further that a plus or minus sign should be placed after a symbol to signify an increase or decrease in length of a particular chromosome or chromosome arm. For example, 46,XY,1q+ symbolizes an increase in the length of the long arm of chromosome 1 of a male having 46 chromosomes, whereas 47,XY,+14p+ symbolizes a male having 47 chromosomes, including an additional chromosome 14 with an increase in length of its short arm.

Refer to Table 10.1 and give the complete descriptive symbols for the chromosome arrangements of the following.

1. A boy having Down syndrome _____

2. A man having Klinefelter syndrome and one Barr body per nucleus _____

3. A woman having Turner syndrome _____

4. A male infant having cri du chat syndrome _____

5. A female newborn having trisomy 18 (Edwards syndrome) _____

6. A male having trisomy 13 (Patau syndrome) _____

IV. PREPARATION OF KARYOTYPES

A karyotype (Figures 10.2, 10.3, and 10.4) is an assemblage of metaphase chromosomes of a particular person arranged in pairs according to order of descending length and position of the centromere. The chromosomes are usually obtained from dividing leukocytes or other kinds of cells in a culture. Cultured cells are treated with colchicine, placed in hypotonic solution, fixed, and stained on microscope slides. With the light microscope, sets of chromosomes from single nuclei are located and evaluated for number of chromosomes present and for proper banding. In medical cytogenetics laboratories, some 10–100 chromosome spreads for each patient are examined through the microscope. The best-appearing chromosome sets are photographed and enlarged to a standard size. Individual chromosomes are then cut out, arranged in matched pairs, and glued in proper order on a standardized form (see page 121). Each karyotype includes a complete set of chromosomes for the person under investigation. Evaluations for structural and numerical chromosome abnormalities are made from the karyotypes prepared for a particular person and verified when necessary from direct microscopic observations of the slides. A karyotype provides a tangible record for comparison, display, and filing for future use.

Today most cytogenetics laboratories use an automated, computer-driven karyotyping system. Although expensive to install, such a system can be accurate and efficient. Because preparing a karyotype in the manner described in this investigation is difficult and time-consuming, automated systems are being used increasingly, especially in major centers where large numbers of karyotypes are prepared each year. Figure 10.6 is an example of a computer-generated karyotype and the chromosome spread from which it was produced. Note that the fetus was found to be triploid ($3n = 69$).

1. To identify chromosomes correctly for a karyotype, proper identification of banding patterns is required. Examine the chromosome spread shown in Figure 10.5, part A, and number several of the chromosomes in the spread. Use the karyotype in Figure 10.5, part B, as a guide to chromosome identification.

2. To familiarize yourself with the banding patterns of all the chromosomes in the human genome, prepare a karyotype from the spread of chromosomes shown in Figure 10.7. This figure is a photograph of a cell of a person having a derivative chromosome. Use a copy of the form "Report of Chromosome Study" (p. 121) for placing the chromosomes in a karyotype, and then answer the following questions concerning the karyotype you have prepared.

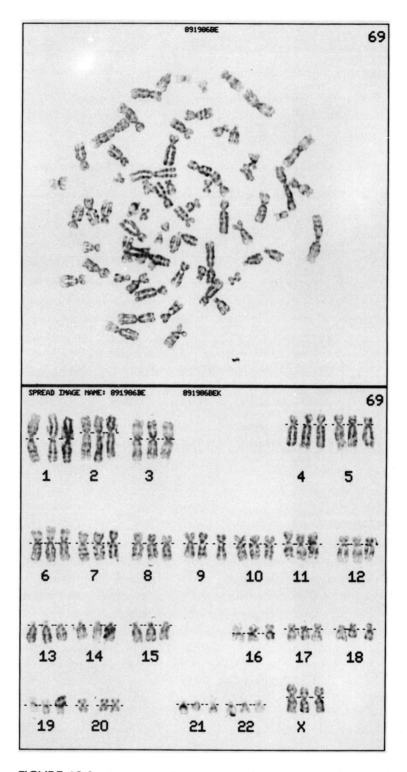

FIGURE 10.6. A computer-driven automated system was used to generate this karyotype of a triploid fetus from the spread of chromosomes in an amniocyte. (Courtesy of Catherine G. Palmer, Indiana University School of Medicine.)

    a. What is the total number of chromosomes in the karyotype? _____

    b. What is the sex of the individual from whom the chromosomes came? _____

    c. Is there a chromosome that does not seem to be matched to any other? _____

    d. If so, compare the banding of the arms of this chromosome with the banding of other chromosomes in the karyotype. Does this comparison allow you to identify the components of this, the derivative chromosome? _____

    e. Do the two arms of the derivative chromosome have banding patterns similar to certain chromosomes in the complement? _____

    f. Would you expect this individual to have a particular syndome? _____ If so, using standard nomenclature, give the descriptive chromosome symbols, sex, and phenotype or syndrome for the karyotype you have prepared. _____

3. To provide all students with an opportunity to gain experience in karyotyping human chromosomes, photographs of chromosome spreads are reproduced for your use as Figures 10.8, 10.9, 10.10, 10.11, 10.12, 10.13, and 10.14. On a copy of the form entitled "Report of Chromosome Study" (page 121), prepare a karyotype of the chromosomes in these figures. Using standard nomenclature, give the descriptive chromosome symbols, sex, and phenotype or syndrome for the karyotype you have prepared.

V. CULTURING AND EXAMINING HUMAN CHROMOSOMES

Many students and instructors might wish to have the experience of actually examining human chromosomes microscopically. Others might even wish to culture human leukocytes, prepare their own microscope slides, photograph chromosome spreads, and prepare karyotypes from these photographs. Prepared slides of human chromosomes are available commercially, and sources for their purchase are listed in the *Instructor's Manual.*

The entire procedure of karyotype preparation is complex and time-consuming and beyond the scope of many general genetics laboratory courses. When time permits and facilities are available, however, students will profit from preparing slides of appropriate cells for chromosome study. For those who have the time and equipment but lack the technical assistance for culturing cells and preparing chromosomes for study, several supply houses market kits with materials and do-it-yourself directions. The supply houses and their products are listed in the *Instructor's Manual.* In addition to the materials provided in the kit, an incubator set at 37°C and a centrifuge are essential to culturing human leukocytes and preparing them for study.

REFERENCES

Bergsma, D., ed. 1972. *Paris Conference (1971): Standardization in human cytogenetics.* Birth Defects Original Article Series, 8(7). White Plains, NY: March of Dimes Birth Defects Foundation.

Borgaonkar, D.S. 1997. *Chromosomal variation in man,* 8th ed. New York: John Wiley & Sons.

Cummings, M.R. 2006. *Human heredity: Principles and issues,* 7th ed. Pacific Grove, CA: Thompson/Brooks/Cole Publishing Co.

deGrouchy, J., and C. Turleau. 1984. *Clinical atlas of human chromosomes,* 2nd ed. New York: John Wiley & Sons.

Gardner, R.J.M., and G.R. Sutherland, 2003. *Chromosome abnormalities and genetic counseling (Oxford monographs on medical genetics, no, 46),* 3rd ed. New York: Oxford University Press.

Hamerton, J.L., and H.P. Klinger, eds. 1975. Paris Conference (1971): Standardization in human cytogenetics, supp. *Cytogenetics and Cell Genetics* 15:201–238.

ISCN. 1995. *An international system for human cytogenetic nomenclature.* Mitelman, F., ed. Basel: S. Karger.

ISCN. 2005. *An international system for human cytogenetic nomenclature.* Shaffer, L.G., and N. Tommerup. eds. Basel: S. Karger.

KEAGLE, M.B., and S.L. GERSEN, eds. 2004. *Principles of clinical cytogenetics*, 2nd ed. Totowa, NJ: Humana Press.

KLUG, W.S., and M.R. CUMMINGS. 2006. *Concepts of genetics.* 8th ed. Upper Saddle River, NJ: Prentice Hall.

LINDSTEN, J.E., H.P. KLINGER, and J.L. HAMERTON. 1978. An international system for human cytogenetic nomenclature. *Cytogenetics and Cell Genetics* 21:311–404.

MCCLATCHEY, K.D., ed. 2002. *Clinical Laboratory Medicine*, 2nd ed. Baltimore: Williams & Wilkins Co.

PAI, G.S., R.C. LEWANDOWSKI, and D.S. BORGAONKAR. 2003. *Handbook of chromosomal syndromes.* New York: John Wiley & Sons.

PATTERSON, D, 1987. The causes of Down syndrome. *Scientific American* 257(8):52–61.

SCHINZEL, A. 2001. *Catalogue of unbalanced chromosome aberrations in man*, 2nd ed. Berlin: Walter de Gruyter.

SPEICHER, M.R., S.G. BALLARD, and D.C. WARD. 1996. Karyotyping human chromosomes by combinational multi-fluor FISH. *Nature Genetics* 12(4): 368–375.

THERMAN, E., and M. SUSMAN. 1993. *Human chromosomes: Structure, behavior, and effects*, 3rd ed. New York: Springer-Verlag.

TIJO, J.H., and A. LEVAN. 1956. The chromosome number of man. *Hereditas* 42:1–6

VERMA, R.S., and A. BABU. 1995. *Human chromosomes—manual of basic techniques*, 2nd ed. Elmsford, NY: Pergamon Press, Inc.

VOGEL, F., and A.G. MOTULSKY. 1996. *Human genetics: Problems and approaches*, 3rd ed. New York: Springer-Verlag.

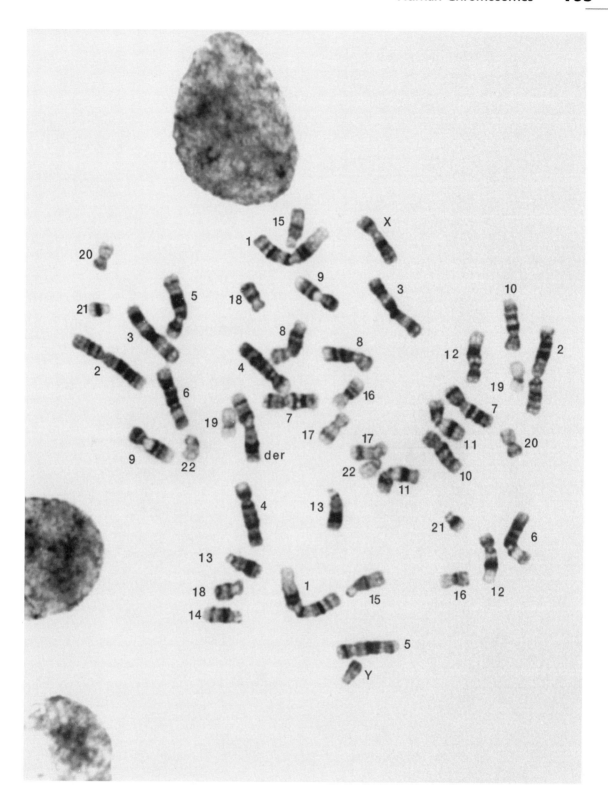

FIGURE 10.7. Photograph illustrating a set of numbered human metaphase chromosomes from a leukocyte culture. Cut out the chromosomes and prepare a karyotype on a copy of the page entitled "Report of Chromosome Study" (page 121). From the information given on the report, answer any questions that you can. Interpret the karyotype and sign the report. This chromosome spread includes a derivative chromosome. Use banding patterns to determine which chromosomes make up the derivative chromosome. (Courtesy of A. Wiktor, J. Zabawski, and D.L. Van Dyke, Department of Medical Genetics, Henry Ford Health System, Detroit, MI.)

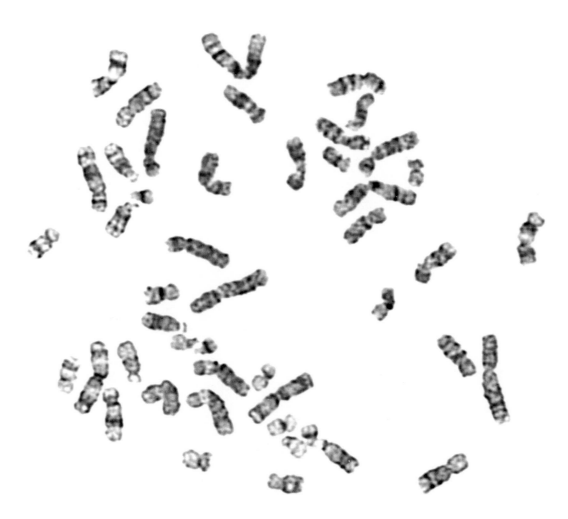

FIGURE 10.8. Set of human metaphase chromosomes from a leukocyte culture. Prepare a karyotype and complete report according to the instructions with Figure 10.7. (Courtesy of D.L. VanDyke, Mayo Clinic).

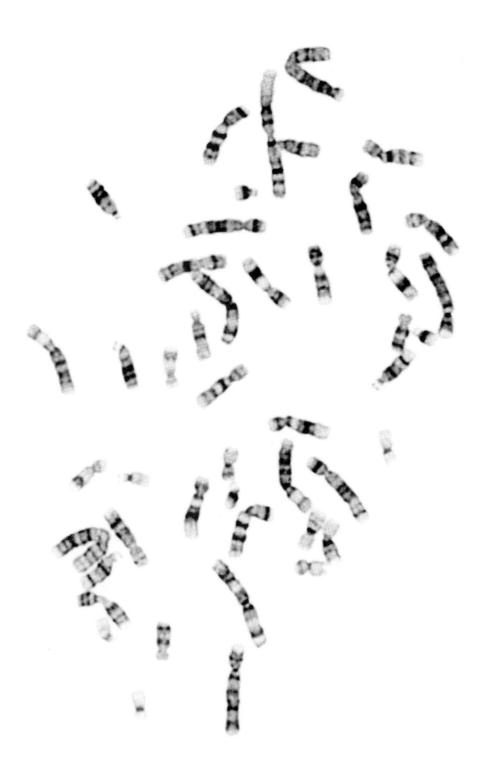

FIGURE 10.9.   Set of human metaphase chromosomes from a leukocyte culture. Prepare a karyotype and complete the report according to the instructions with Figure 10.7. (Courtesy of D.L. Van Dyke and J. Zabawski, Henry Ford Hospital, Detroit.)

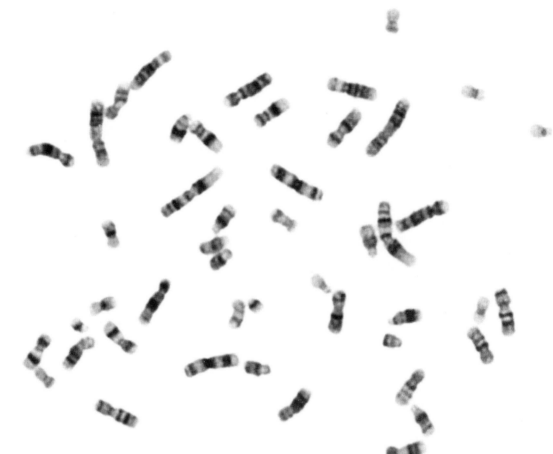

FIGURE 10.10. Set of human metaphase chromosomes from a leukocyte culture. Prepare a karyotype and complete the report according to the instructions with Figure 10.7. (Courtesy of D.L. Van Dyke and J. Zabawski, Henry Ford Hospital, Detroit.)

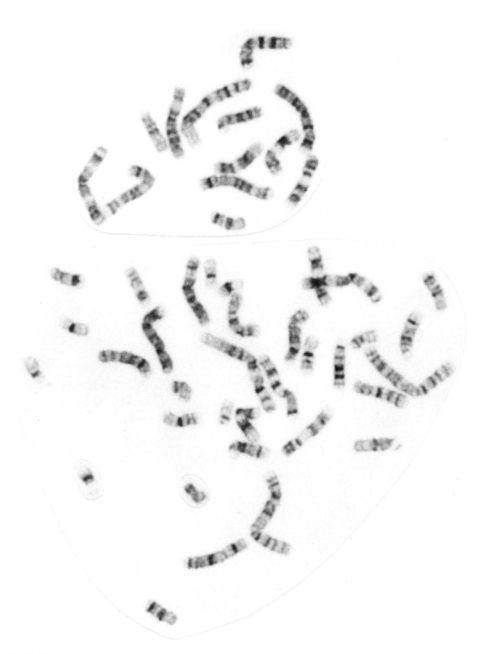

FIGURE 10.11.   Set of human metaphase chromosomes from a leukocyte culture. Prepare a karyotype and complete the report according to the instructions with Figure 10.7. (Courtesy of D.L. Van Dyke.)

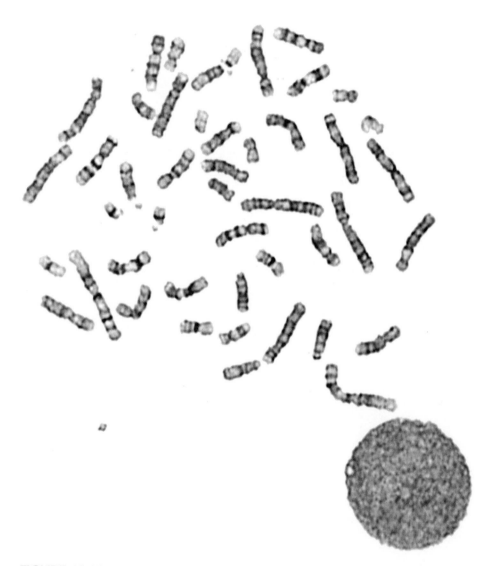

FIGURE 10.12.   Set of human metaphase chromosomes from a leukocyte culture. Prepare a karyotype and complete report according to the instructions with Figure 10.7. (Courtesy of D.L. VanDyke, Mayo Clinic).

FIGURE 10.13. Set of human metaphase chromosomes from a leukocyte culture. Prepare a karyotype and complete the report according to the instructions with Figure 10.7. (Courtesy of D.L. Van Dyke and J. Zabawski, Henry Ford Hospital, Detroit.)

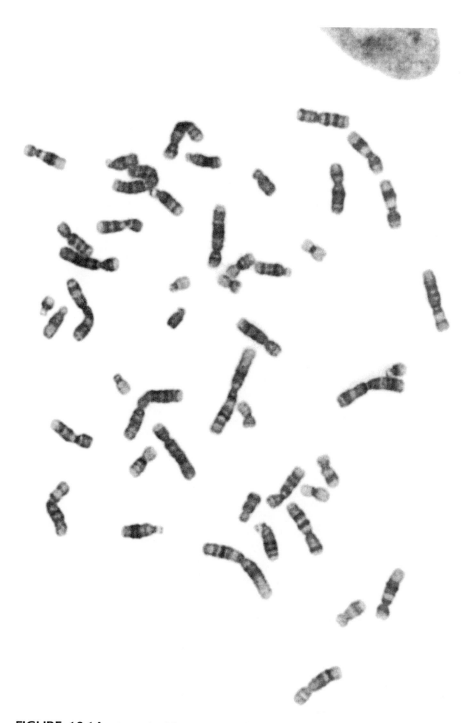

FIGURE 10.14.   Spread of human metaphase chromosomes from a leukocyte culture. Prepare a karyotype and complete the report according to the instructions with Figure 10.7. (Courtesy of A. Wiktor, J. Zabawski, and D.L. Van Dyke, Department of Medical Genetics, Henry Ford Health System, Detroit, MI.)

Report of Chromosome Study

Cytogenetic Laboratory

1	2	3	4	5
	A Group			B Group

6	7	8	9	10	11	12
			C Group			

13	14	15	16	17	18
	D Group			E Group	

19	20	21	22	Y	
	F Group			G Group	X

Name   Unit No.

Culture No.   Referred by

Date of Culture   Ward or Clinic

Chromosome Counts 45 46 47 Other Total

No. of Cells

Karyotypes

Interpretation

Signed _____

INVESTIGATION 11

Linkage and Crossing Over

As was noted in Investigation 1, had Mendel investigated additional gene loci in garden peas, he no doubt would have encountered the phenomenon of linkage. When other scientists eventually studied the genetics of various organisms with thoroughness, they inevitably found exceptions to Mendel's law of independent assortment.

In fact, as early as 1903, Walter S. Sutton, an American geneticist, proposed that gene behavior could be understood best if it were assumed that the genes are associated with the chromosomes. Sutton also suggested that in most organisms the number of genes far exceeds the number of chromosomes, which means that genes on the same chromosome would not be expected to assort independently but would be linked to one another. In 1905, Bateson, Saunders, and Punnett reported the first example of nonindependent assortment. Using sweet peas, they demonstrated that a gene affecting flower color and one determining the shape of the pollen grain did not behave according to Mendel's second law. Unfortunately, they did not interpret their data in terms of linkage but suggested an alternative (and erroneous) hypothesis.

As linkage was studied further, it was found not to be absolute. That is, recombination occurs between linked genes by the process of crossing over (the reciprocal exchange of chromosomal material between homologous chromosomes). The amount of recombination varies with the distance between the genes: The greater the distance, the higher the percentage of recombination. By combining the information about linkage as it affects a number of genes, linkage maps have been documented for many species. In these maps, large numbers of genes are related to one another on the basis of the percentage of crossing over between them. Especially detailed linkage maps were developed for *Drosophila*, maize, and a number of bacteria and viruses.

The following hypothetical example illustrates how crossover percentages may be employed in mapping genes on chromosomes. Assume that three two-point (dihybrid) crosses are performed.

$$\frac{ab}{ab} \times \frac{AB}{AB} \rightarrow F_1 \frac{AB}{ab}; \text{ testcross } F_1 \text{ to } \frac{ab}{ab} \rightarrow 6\% \text{ crossing over between } A \text{ and } B$$

$$\frac{bc}{bc} \times \frac{BC}{BC} \rightarrow F_1 \frac{BC}{bc}; \text{ testcross } F_1 \text{ to } \frac{bc}{bc} \rightarrow 10\% \text{ crossing over between } B \text{ and } C$$

$$\frac{ac}{ac} \times \frac{AC}{AC} \rightarrow F_1 \frac{AC}{ac}; \text{ testcross } F_1 \text{ to } \frac{ac}{ac} \rightarrow 15\% \text{ crossing over between } A \text{ and } C$$

On the basis of these data from three crosses and three testcrosses, the genes can be mapped as follows:

$$\overline{\quad A \quad 6 \quad B \quad 10 \quad C \quad}$$
$$\longleftarrow\!\!\!-15-\!\!\!\longrightarrow$$

Thus, both gene order (*ABC*) and the distance between the genes can be established on the basis of recombination frequencies. One genetic **map unit** is defined as 1% of recombination between two genes. Thus, *A* and *B* can be said to be six map units apart.

The discrepancy between the sum of the intermediary distances (6% + 10%) and the distance (15%) between the end genes can be explained by further experiments. A three-point (trihybrid) cross, using one cross followed by a testcross, will provide more facts and solve the dilemma of $6 + 10 \neq 15$.

$$\frac{abc}{abc} \times \frac{ABC}{ABC} \rightarrow F_1 \frac{ABC}{abc}; \text{ testcross the } F_1 \text{ to } \frac{abc}{abc} \text{ to yield the following data:}$$

	Phenotype	Number
Parental offspring	*ABC*	862
	abc	828
Single crossovers between *A* and *B*	*Abc*	51
	aBC	59
Single crossovers between *B* and *C*	*ABc*	102
	abC	88
Double Crossovers	*AbC*	6
	aBc	4
		2000

Using these testcross data, you can calculate the percentages of crossing over between *A* and *B*, *B* and *C*, and *A* and *C*. In making these calculations, add *all* of the crossovers that take place between any two genes and divide by the total number of progeny.

$$A - B: \quad \frac{51 + 59 + 6 + 4}{2000} = \frac{120}{2000} = 6\%$$

$$B - C: \quad \frac{102 + 88 + 6 + 4}{2000} = \frac{200}{2000} = 10\%$$

$$A - C: \quad \frac{51 + 59 + 102 + 88 + 2(6 + 4)}{2000} = \frac{320}{2000} = 16\%$$

Note that these calculations give the percentage of crossing over between the end genes (*A* and *C*) as 16%, which does not agree with the percentage of recombination (15%) obtained in the dihybrid cross. A two-point cross, however, does not detect **double-crossover** individuals; such individuals are identical in phenotype to the parental type of progeny and would be classified with them in the two-point cross. The three-point cross gives you the advantage of being able to recognize double-crossover progeny produced by simultaneous crossovers between *A* and *B* and *B* and *C*. If you calculate the **percentage of recombination** between the end genes (*A* and *C*) from the three-point data, the double-crossover individuals are again classified as parental types (with respect to *A* and *C*) and you find

$$\text{percentage of recombination} = \frac{51 + 59 + 102 + 88}{2000} = \frac{300}{2000} = 15\%$$

One additional statistic can be measured from three-point linkage data, the **coefficient of coincidence**. This statistic is defined as the **observed number of double-crossover individuals divided by the expected number of double-crossover individuals**. The expected number of double-crossover individuals can be calculated on the basis of the principle that the probability of two independent events occurring simultaneously equals the product of their individual probabilities. Thus, because the probability of crossing over between A and B is 6%, and that between B and C is 10%, it follows that the probability of double crossing over is $(.06) \times (.10) = .006$. Multiplying .006 times 2000, the total number of progeny in the cross, yields 12, the expected number of double crossovers. Therefore, the coefficient of coincidence $= 10/12 = 5/6 = 0.8333$, or 83.33% This means that there is some **interference**; that is, a crossover in one region ($A - B$) of the chromosome interferes with a crossover in the adjacent region ($B - C$), or vice versa. A coefficient of coincidence of 0 would mean **complete** interference, whereas a coefficient of 1 would mean no interference. Generally, for *Drosophila*, one obtains coefficients of coincidence of greater than 0 but less than 1. This means there is **some** but not **complete** interference.

In addition to determining recombination frequencies and genetic map units, three-point testcrosses permit one to determine the correct order of the three genes on the chromosome. Consult a genetics textbook for further discussion on how this is done.

In this investigation, you will conduct trihybrid crosses with *Drosophila*. Using the data obtained, you will determine gene order and map the genes on the chromosome. In addition, since conducting three-point testcrosses would be very difficult in humans, several procedures for locating and mapping genes in humans are discussed.

OBJECTIVES OF THE INVESTIGATION

Upon completion of this investigation, you should be able to

1. **analyze** three-point linkage data and map the genes on the chromosome,
2. **measure** the map distance between autosomally linked genes using dihybrid F_2 data and a statistical table designed to permit achieving this objective,
3. **describe** how mouse–human somatic cell hybridization techniques can be used to associate a particular human gene with a particular chromosome, and
4. **describe** how fluorescent in *situ* hybridization (FISH) and chromosome painting are used to locate specific DNA sequences on particular human chromosomes and to analyze complex chromosome rearrangements.

Materials needed for this investigation:

    materials for handling *Drosophila* (see Investigation 1)

    Drosophila stocks: wild type; mutant stock such as yellow body, forked bristles, miniature wings

      ($y f m$); other stock or stocks selected by the instructor and designated as _____

I. GENE MAPPING IN *DROSOPHILA*

A. Procedure

Linkage may be measured for either X-linked or autosomal genes. In the former case, if a multiple-mutant female is mated with a wild-type male to produce an F_1, one can subsequently produce an F_2 in which crossover percentages can be readily measured. By contrast, if the genes are autosomal, one must testcross an F_1 female to determine crossover percentages conveniently.

1. Why does one use different procedures to measure linkage for X-linked and autosomal genes? (Try diagramming a real or hypothesized cross, first with X-linked and then with autosomal genes; then, write a brief answer to this question.) _____

_____

_____

_____

2. Now mate the fly stocks provided by the instructor. For example, you might be asked to mate a virgin _y f m_ female with a wild-type male. Produce the F_1 flies and determine whether the genes involved are autosomal or X-linked. If _y_, _f_, and _m_ are X-linked, how will F_1 phenotype(s) differ from what would be produced if the genes were autosomal? _____

_____

_____

_____ _____

3. If the genes are autosomal, testcross virgin F_1 females with multiple-mutant males. Record all data relative to this cross and the testcross in Table 11.1. Also perform the **reciprocal cross**; that is, mate an F_1 male with a virgin, multiple-mutant female. Record data from this experiment in Table 11.2. If the genes are X-linked, mate an F_1 female with an F_1 male, produce an F_2 generation, and record all data for this experiment in Table 11.1.

B.   Evaluation of Results

On the basis of the results obtained in this experiment, answer the following questions.

1. Did you find the genes involved in the experiment you performed to be autosomal or X-linked?

_____

2. If the genes were autosomal, how did the results of the reciprocal testcrosses differ from one

another? _____

_____

What conclusion does this difference lead you to make? _____

_____

_____

3. Using the total number of testcross or F_2 progeny from Table 11.1, answer the following questions:

     a. What is the correct **gene order** for the three genes studied in the experiment? _____

     b. Showing the genes on chromosomes, diagram the **genotypes** of the following.

         P_1 female _____

         P_1 male _____

         F_1 female _____

         F_1 male _____

TABLE 11.1. Record of a *Drosophila* Experiment

Experiment involving three-point linkage. Use this table if the genes are autosomal (testcross an F_1 female) or if the genes are X-linked (produce an F_2)

Experiment number _____ **Name** _____

1. Cross _____ female × _____ male

2. Date P_1s mated: _____

3. Date P_1s removed: _____

4. Date F_1s first appeared: _____

5. Phenotype of F_1 males: _____

6. Phenotype of F_1 females: _____

7. Date F_1 male and female placed in fresh bottle (or date F_1 virgin female testcrossed): _____

8. Date F_1 flies removed: _____

9. Date F_2 (or testcross) progeny appeared: _____

10. Record F_2 (or testcross) data in the following table:

	Males		Females		
	Phenotype	Number	Phenotype	Number	Total
a.	_____	_____	_____	_____	_____
b.	_____	_____	_____	_____	_____
c.	_____	_____	_____	_____	_____
d.	_____	_____	_____	_____	_____
e.	_____	_____	_____	_____	_____
f.	_____	_____	_____	_____	_____
g.	_____	_____	_____	_____	_____
h.	_____	_____	_____	_____	_____
Totals		_____		_____	_____

TABLE 11.2. Record of a *Drosophila* Experiment

Reciprocal testcross data for experiment involving autosomal genes. Use this table for data from a testcross of an F_1 male.

Experiment number _____ **Name** _____

1. Cross _____ female × _____ male

2. Date P_1s mated: _____

3. Date P_1s removed: _____

4. Date F_1s first appeared: _____

5. Phenotype of F_1 males: _____

6. Phenotype of F_1 females: _____

7. Date F_1 male was testcrossed: _____

8. Date F_1 male and its mate are removed: _____

9. Date testcross flies appeared: _____

10. Record testcross data in the following table:

	Males		Females		
	Phenotype	Number	Phenotype	Number	Total
a.	_____	_____	_____	_____	_____
b.	_____	_____	_____	_____	_____
c.	_____	_____	_____	_____	_____
d.	_____	_____	_____	_____	_____
e.	_____	_____	_____	_____	_____
f.	_____	_____	_____	_____	_____
g.	_____	_____	_____	_____	_____
h.	_____	_____	_____	_____	_____
Totals		_____		_____	_____

c. Calculate the **percentage of crossing over** between the following gene pairs (designate the symbols):

Symbols   % Crossing Over

_____ and _____   :_____

_____ and _____   :_____

_____ and _____ (the end genes):_____

d. Calculate the **percentage of recombination** between the end genes (i.e., exclude double crossovers). _____

e. Calculate the expected number of double-crossover individuals. _____

_____

f. Calculate the coefficient of coincidence. _____

_____

g. Interpret the meaning of this coefficient of coincidence. _____

_____

4. In the charts that follow, calculate chi-square (χ^2) for each of the three traits by themselves. What is the expected ratio of dominant to recessive individuals for each trait?

_____

_____

Use this theoretical expectation to calculate χ^2.

Trait 1

Phenotype	O	E	O − E	$(O - E)^2$	$(O - E)^2/E$
_____	_____	_____	_____	_____	_____
_____	_____	_____	_____	_____	_____
Total	_____			$\chi^2 =$	_____
				$P =$	_____

Conclusion: I **accept/reject** the hypothesis that the data fit the expected ratio.

Trait 2

Phenotype	O	E	O − E	$(O - E)^2$	$(O - E)^2/E$
_____	_____	_____	_____	_____	_____
_____	_____	_____	_____	_____	_____
Total	_____			$\chi^2 =$	_____
				$P =$	_____

Conclusion: I **accept/reject** the hypothesis that the data fit the expected ratio.

Trait 3

Phenotype	O	E	O − E	$(O - E)^2$	$(O - E)^2/E$
Total	_____			$\chi^2 = $ _____	
				$P = $ _____	

Conclusion: I **accept/reject** the hypothesis that the data fit the expected ratio.

5. In the following chart, calculate χ^2 on the basis of the hypothesis that the three traits are independently assorting. If they do assort independently, what is the expected ratio?

_____

_____

Use the total of males and females from Table 11.1 for this calculation.

Phenotype	O	E	O − E	$(O - E)^2$	$(O - E)^2/E$
Total	_____			$\chi^2 = $ _____	
				$P = $ _____	

Conclusion: I **accept/reject** the hypothesis that the data fit the expected ratio.

II. MEASURING LINKAGE IN F₂ POPULATIONS

Determining the percentage of crossingover between autosomally linked genes is easiest to accomplish in testcross populations. However, with the aid of Table 11.3, one can measure linkage in F_2 populations of a dihybrid cross.

TABLE 11.3. Table for Calculating Linkage Intensities (Percentages of Crossing Over) from F_2 Data*

	Ratio of Products			Ratio of Products			Ratio of Products	
Crossover Value	$\dfrac{ad}{bc}$ (Repulsion)	$\dfrac{bc}{ad}$ (Coupling)	Crossover Value	$\dfrac{ad}{bc}$ (Repulsion)	$\dfrac{bc}{ad}$ (Coupling)	Crossover Value	$\dfrac{ad}{bc}$ (Repulsion)	$\dfrac{bc}{ad}$ (coupling)
.005	.00005000	.00003361	.205	.09351	.08140	.405	.5079	.5007
.010	.00020005	.0001356	.210	.09865	.08628	.410	.5266	.5199
.015	.0004503	.0003076	.215	.1040	.09136	.415	.5460	.5398
.020	.0008008	.0005516	.220	.1095	.09663	.420	.5660	.5603
.025	.001252	.0008692	.225	.1152	.1021	.425	.5867	.5815
.030	.001804	.001262	.230	.1211	.1078	.430	.6081	.6034
.035	.002458	.001733	.235	.1272	.1137	.435	.6302	.6260
.040	.003213	.002283	.240	.1334	.1198	.440	.6531	.6494
.045	.004070	.002914	.245	.1400	.1262	.445	.6768	.6735
.050	.005031	.003629	.250	.1467	.1328	.450	.7013	.6985
.055	.006096	.004429	.255	.1536	.1396	.455	.7266	.7243
.060	.007265	.005318	.260	.1608	.1467	.460	.7529	.7510
.065	.008540	.006296	.265	.1682	.1540	.465	.7801	.7786
.070	.009921	.007366	.270	.1758	.1616	.470	.8082	.8071
.075	.01141	.008531	.275	.1837	.1695	.475	.8374	.8366
.080	.01301	.009793	.280	.1919	.1777	.480	.8676	.8671
.085	.01471	.01116	.285	.2003	.1861	.485	.8990	.8986
.090	.01653	.01262	.290	.2089	.1948	.490	.9314	.9313
.095	.01846	.01419	.295	.2179	.2038	.495	.9651	.9651
.100	.02051	.01586	.300	.2271	.2132	.500	1.0000	1.0000
.105	.02267	.01765	.305	.2367	.2228	.505	1.0362	1.0362
.110	.02495	.01954	.310	.2465	.2328	.510	1.0738	1.0736
.115	.02734	.02156	.315	.2567	.2432	.515	1.1128	1.1124
.120	.02986	.02375	.320	.2672	.2538	.520	1.1533	1.1526
.125	.03250	.02594	.325	.2780	.2649	.525	1.1953	1.1942
.130	.03527	.02832	.330	.2899	.2763	.530	1.2390	1.2373
.135	.03816	.03083	.335	.3008	.2881	.535	1.2844	1.2819
.140	.04118	.03347	.340	.3127	.3002	.540	1.3316	1.3282
.145	.04434	.03624	.345	.3250	.3128	.545	1.3806	1.3762
.150	.04763	.03915	.350	.3377	.3259	.550	1.4317	1.4260
.155	.05105	.04220	.355	.3508	.3393	.555	1.4847	1.4776
.160	.05462	.04540	.360	.3643	.3532	.560	1.5400	1.5312
.165	.05832	.04875	.365	.3783	.3675	.565	1.5975	1.5868
.170	.06218	.05240	.370	.3927	.3823	.570	1.6574	1.6446
.175	.06618	.05591	.375	.4076	.3977	.575	1.7198	1.7045
.180	.07033	.05972	.380	.4230	.4135	.580	1.7848	1.7668
.185	.07464	.06371	.385	.4389	.4298	.585	1.8526	1.8316
.190	.07911	.06787	.390	.4553	.4467	.590	1.9234	1.8989
.195	.08374	.07220	.395	.4723	.4641	.595	1.9972	1.9689
.200	.08854	.07670	.400	.4898	.4821	.600	2.0742	2.0417

*Let the four observed F_2 classes be represented by

AB	Ab	aB	ab	Total
(a)	(b)	(c)	(d)	N

For repulsion, calculate ad/bc and for coupling, bc/ad. Determine the crossover value (expressed as a decimal fraction) by interpolation in this table.

Source: Table adapted from R.A. Fisher and B. Balmukand, The estimation of linkage from the offspring of selfed heterozygotes. *Journal of Genetics* 20(1928–1929):79–92. Indian Academy of Sciences, Bangalore.

Mertens and Burdick (1954) reported their studies of inheritance of two genes, f and bi, in the tomato, *Lycopersicon esculentum*. The gene f, fasciated, increases the number of locules (compartments) in the fruits and results in fruits having a flattened, irregular outline. The gene bi, bifurcate, causes the inflorescence to be divided into two distinct branches. Mertens and Burdick reported the following experimental cross:

$$P_1 \frac{++}{++} \times \frac{bi\,f}{bi\,f} \rightarrow F_1 \frac{++}{bi\,f}$$

The phenotypically wild-type F_1 plants were allowed to self-fertilize and the following F_2 data were collected.

Category	Phenotype	Observed number	Expected number
(a)	++	133	108.5625
(b)	bi+	4	36.1875
(c)	+f	4	36.1875
(d)	bif	52	12.0625
Totals		193	193.0000

The expected numbers shown are based on the assumptiom of independent assortment (9:3:3:1 ratio) between bi and f. Chi-square was calculated to be 194.99 with 3 degrees of freedom. The probability that the deviations between the expected and the observed were due to chance alone is thus much less than 0.01, resulting in the rejection of the hypothesis of independent assortment of bi and f. Therefore, bi and f most probably are linked with a measurable amount of crossing over.

Using Table 11.3, we can calculate the statistic bc/ad as follows, since the genes are in coupling (cis) phase linkage (*note: b, c, a,* and *d* signify the "categories" listed in the data given).

$$\frac{bc}{ad} = \frac{4 \times 4}{133 \times 52} = \frac{16}{6916} = 0.002313$$

Now, looking in the coupling column of Table 11.3, we find the calculated value of bc/ad to lie between the crossover values of .040 and .045. We can use table values to interpolate and thereby calculate the map distance between the genes bi and f, as follows:

Map distances in decimal fractions	Values of bc/ad
.040	.002283
x	.002313
.045	.002914

$$\frac{x - .040}{.045 - .040} = \frac{.002313 - .002283}{.002914 - .002283}$$

$$\frac{x - .040}{.005} = \frac{.00003}{.000631}$$

$$x - .040 = \frac{.00003 \times .005}{.000631} = .000238$$

$$x = .040 + .000238 = .040238$$

For all practical purposes, then, the crossover distance between bi and f is four map units.

Now, consider the following data provided by the late Eldon J. Gardner of Utah State University. A cross was made between two true-breeding varieties of barley. One variety had purple achenes and two-rowed heads. The other variety was yellow and six-rowed. The F_1 plants were purple, two-rowed. The entire F_2 population was classified as follows.

Phenotype	Number
(*a*) Purple, two-rowed	3428
(*b*) Purple, six-rowed	225
(*c*) Yellow, two-rowed	286
(*d*) Yellow, six-rowed	1118

1. What F_2 results would you expect on the hypothesis of independent assortment? _____

    _____

    _____

2. Complete the following chart, calculate χ^2, and determine whether the hypothesis of independent assortment is acceptable.

    _____

Phenotype	O	E	O − E	$(O - E)^2$	$(O - E)^2/E$
Purple, two-rowed	3428	_____	_____	_____	_____
Purple, six-rowed	225	_____	_____	_____	_____
Yellow, two-rowed	286	_____	_____	_____	_____
Yellow, six-rowed	1118	_____	_____	_____	_____
Total	_____			$\chi^2 =$	_____
				$P =$	_____

Conclusion: I **accept/reject** the hypothesis of independent assortment.

3. Is linkage present? _____

    What evidence supports your answer? _____

    _____

4. The gene controlling the arrangement of achenes, that is, two-rowed or six-rowed, may be symbolized by *T* or *t* and the gene for color by *P* or *p*. On the basis of the hypothesis that linkage is present, give the following genotypes by showing the genes on chromosomes.

    a. Purple, two-rowed parent: _____

   b. Yellow, six-rowed parent: _____

   c. Purple, two-rowed F_1: _____

   d. Purple, two-rowed F_2 s: _____

   e. Purple, six-rowed F_2 s: _____

   f. Yellow, two-rowed F_2 s: _____

   g. Yellow, six-rowed F_2 s: _____

5. Are the genes in the F_1 in coupling *PT/pt* or repulsion *Pt/pT* phase of linkage? _____

   _____

6. Use Table 11.3 to calculate the percentage of recombination between the genes. The numerical
    values for *a, b, c,* and *d* can be found on page 133 where the F_2 data are first given.
    a. Calculate *ad/bc* (repulsion) or *bc/ad* (coupling) as the case demands; record the value here.

    _____

    b. Determine the crossover value (expressed as a decimal fraction) by interpolation in Table 11.3.
        Record the value here.

    _____

III. MAPPING GENES ON HUMAN CHROMOSOMES

Geneticists have developed extensive linkage maps for such organisms as *Drosophila*, maize, certain
viruses and bacteria, and the mouse. By contrast, until recently, progress in preparing human linkage
maps has been slow. The reasons for this are fairly obvious. (1) Geneticists cannot carry out controlled
matings using humans. (2) Human families are relatively small and thus provide geneticists with too
few individuals for measuring linkage relationships adequately. (3) The human generation time is
long. (4) Because most abnormal genetic traits are rare, it is an extremely rare pedigree in which two
such traits (which may or may not be linked) happen to occur in the same individual. For example, it
is rare for two X-linked recessive genes, such as those for hemophilia and color blindness, to occur in
the same individual. To measure linkage intensities for X-linked genes using traditional pedigree techniques requires a study of the male children of women who are heterozygous for two such genes.
Thus, if a woman has inherited the gene for color blindness from one parent and the gene for hemophilia from the other, her sons who express both mutant traits or neither mutant trait bear crossover
X chromosomes. Establishing the linkage of autosomal genes and measuring the intensity of that linkage
are even more difficult.

   Although the techniques of classical genetics have allowed for some linkage relationships to be
established in humans, extensive mapping of human chromosomes is dependent on such special techniques as those employed in somatic cell hybridization [see McKusick (1971, 1980) and Ruddle and
Kucherlapati (1974)] and *in situ* hybridization of DNA probes to chromosomes. According to online
Mendelian Inheritance in Man (OMIM),[1] by October 2005 some 16,306 human gene loci were known,
including 15,279 autosomal, 909 X-linked, and 56 Y-linked. In addition to the genes assigned to the X
and Y chromosomes, 8797 autosomal genes have been assigned to specific chromosomes, with many

---

[1]   Updated information on the number of human genes and the Human Genome Project can be obtained on the
      net at OMIM (Online Mendelian Inheritance in Man, V.A. McKusick, ed.) http://www.ncbi.nlm.nih.gov/omim/

located at specific sites within their respective chromosomes. In December 1999, it was announced that chromosome 22 had been completely sequenced. This was followed in May 2000 with the announcement that chromosome 21 had also been completely sequenced. The rest of the human genome was completely sequenced in 2003.

Subsequent research has concentrated on assigning specific functions to the genes encoded by the estimated 3 billion base pairs in the human genome. This research will, of course, allow definitive assignment of all human genes to specific chromosomes and locations.

A.   Somatic Cell Hybridization

Ruddle and Kucherlapati (1974) discussed interspecific somatic cell hybridization experiments, notably, mouse–human hybrids, that have accounted for much of the early progress in mapping genes to specific chromosomes. How mouse–human hybrid somatic cells cultured *in vitro* can be used to assign genes to specific chromosomes is especially interesting. Spontaneous fusion of mouse and human cells growing together in tissue culture will occur with a low frequency. Such fusion may be facilitated considerably by the addition of inactivated Sendai viruses or polyethylene glycol, which stimulates the formation of intercellular bridges between adjacent cells. Not only do the cells fuse but their nuclei also fuse. As such heterokaryotic cells divide mitotically, the 40 mouse chromosomes generally are retained intact, whereas the human chromosomes are randomly and preferentially lost. Thus, a hybrid somatic cell usually does not contain 86 chromosomes (46 human + 40 mouse) but generally does contain 40 mouse chromosomes plus a few human chromosomes. Ruddle and Kucherlapati claim that such hybrid cells usually contain a total of 41−55 chromosomes.

Some of the genes on both mouse and human chromosomes in hybrid somatic cells are capable of expressing themselves simultaneously *in vitro*. Mouse genes code uniquely mouse proteins, and human genes code human proteins. Electrophoretic procedures can be used to distinguish mouse and human protein products from one another.

Through banding techniques for staining (see Investigation 10), mouse and human chromosomes can be distinguished and specific human chromosomes present in the cells can be identified. Thus, if a particular hybrid cell line with 40 mouse chromosomes and 1 human chromosome can be shown to produce a certain human enzyme, it follows that the gene coding for that particular enzyme must be located in the particular human chromosome present in the hybrid cell line.

This procedure for assigning a particular gene to a particular chromosome is not always as simple as the last few paragraphs indicate. Unfortunately, not all genes will express themselves when the cells containing them are grown in tissue culture. Such genes cannot be detected *in vitro*. In addition, many of the mouse–human hybrid clones contain several human chromosomes rather than just one. As a result, it is more difficult to associate a particular gene product with a particular chromosome. Despite the technical difficulties and the laboriousness of the procedure, at one time somatic cell hybridization experiments had resulted in assigning more genes to specific chromosomes in humans than any other technique (McKusick, 1998). Because of new technical developments (see Section IIIB, below, and Investigation 15), this statement is no longer true.

Study Table 11.4 and answer the questions relative to it. This exercise will help you to understand how, by the process of elimination, the genes that produce specific gene products can be associated with particular chromosomes.

You should interpret the data in Table 11.4 as follows: Hybrid cell line I produces gene products A, B, C, and E but not D. Cell line I has two human chromosomes, numbers 3 and 18, from among the four possibilities indicated. Now, answer the following questions:

1. Assuming that each gene product is controlled by a different gene, can you determine whether any of the five gene products—A, B, C, D or E—are controlled by genes linked on the same chromosome?

_____ If so, which gene products are controlled by linked genes? _____

TABLE 11.4. Hypothetical Data Illustrating How Certain Gene Products (Enzymes or Other Proteins) Can Be Associated with Certain Human Chromosomes in Mouse–Human Hybrid Cell Lines*

	Mouse–Human Hybrid Cell Line				
	I	II	III	IV	V
Human gene products					
A	+	+	−	−	−
B	+	−	+	−	−
C	+	+	−	−	−
D	−	−	+	+	+
E	+	−	+	−	−
Human chromosomes					
3	+	−	+	−	−
12	−	−	+	+	+
18	+	+	−	−	−
21	−	−	+	−	−

*A plus (+) indicates the presence of a particular human protein (enzyme) or a particular chromosome, and a minus (−) indicates the absence of the enzyme or chromosome in the hybrid clone.

What is the evidence for this conclusion? _____

_____

_____

2. Can you determine on which chromosome(s) the linked genes are located? _____

   Identify the chromosome(s) by giving its number(s). _____

   What evidence supports this conclusion? _____

   _____

   _____

3. Is it possible to associate any other gene or genes with a particular chromosome?_____

   If so, which gene on which chromosome? _____

   What is the evidence for this conclusion?_____

   _____

   _____

   _____

4. Are any of the chromosomes listed in Table 11.4 not associated with any of the gene products

   A−E? _____

B. Molecular Techniques: *In Situ* Hybridization

In recent years molecular techniques have been developed for mapping genes and other types of DNA sequences to specific chromosomes. Although it is beyond the scope of this investigation, one of the most valuable techniques used in this type of mapping is **in situ hybridization**.

With *in situ* hybridization, any isolated DNA sequence or gene can be mapped to a specific human chromosome if humans possess that gene or have a very similar type of gene. In *Drosophila*, for example, homeobox genes, which are regulatory developmental genes (genes that switch on whole groups of genes expressed temporally during development), were initially characterized by classical mutational analysis. These mutants, which might, for example, cause an antenna to develop into a leg (*Antennapedia*) or the duplication of a whole segmental region, as is the case with *Ultrabithorax* (two thorax-like regions instead of one), ultimately led to the isolation of the DNA genes that, when altered, were responsible for these mutant *Drosophila* phenotypes. Once these *Drosophila* genes were isolated, it was possible to demonstrate (via cross hybridization) that other organisms, including humans, also possessed similar genes. For mammals these genes are referred to as the *Hox* genes (actually clusters of genes) and are thought to be involved in the establishment of the anterior-posterior axis during development. *In situ* hybridization has allowed the mapping of these genes to specific human chromosomes. For example, the gene for the *Drosophila* mutant *Antennapedia* (*Antp*) shows great similarity to the *Hox A6*, *Hox B6*, and *Hox C6* genes found in humans. *In situ* hybridization has established that *Hox B6* maps to the human chromosome 17q21 area and *Hox C6* to the chromosome 12q12 area. A similar approach has led to the localization of many other human genes, including oncogenes (cancer genes).

The procedure for *in situ* hybridization involves growing cells such as leukocytes in culture and preparing metaphase or prometaphase chromosome spreads as described in Investigation 10. The chromosome spreads are then treated with the enzymes RNase and Proteinase K to remove chromosomal proteins and RNA and to expose "naked" chromosomal DNA. The slide with the chromosome spreads is subjected to DNA denaturing solution, which serves to separate the two complementary strands of the DNA double helix, rendering the chromosomal DNA single-stranded. The chromosome spread is covered with a solution containing the isolated **DNA probe** of interest (for example, isolated *Drosophila Antennapedia* DNA), which has been "tagged" either with the radioactive isotope tritium or with modified nucleotides and also made single-stranded and then subjected to renaturation conditions. Any human chromosome that possesses a DNA sequence complementary to that of the DNA-probe will bind the probe during this renaturation period to form a chromosome DNA-probe DNA double helix duplex. The chromosome preparation is then washed free of any unbound DNA probe and the bound DNA probe visualized either by autoradiography (localization of the tritium) or, if modified nucleotides were used to tag the probe, by treatment with fluorescent dyes, which are visualized with ultraviolet light. Any place on the human chromosomes at which there is a specific binding of the DNA probe indicates where the human gene (or at least a similar gene) of that type is located. This latter process of using modified nucleotides and fluorescent dyes is more precise and is termed *fluorescent in situ hybridization (FISH)*. The FISH procedure is now widely used in the mapping of specific DNA sequences and in a procedure termed **chromosome painting**, in which a mixture of different chromosome-specific probes tagged with different fluorescent dyes is hybridized simultaneously to metaphase chromosomes, the effect being to paint the various chromosomes in different colors. Chromosome painting is also becoming very useful in karyotyping chromosomes and observing chromosome rearrangements (see Investigation 10).

The color photograph on the back cover of this manual is of a human leukocyte from a patient with chronic granulocytic leukemia. The metaphase chromosome spread has been prepared using the FISH procedure to differentially paint the chromosomes. (Note that chromosome 22 has been painted to fluoresce pink and the other chromosomes blue.) This particular type of leukemia is associated with specific chromosome aberrations involving chromosomes 9, 12, and 22. Officially, these aberrations involve a translocation between chromosomes 9 and 22 creating a derived (der) chromosome 9,22 and a very short Philadelphia chromosome (Ph′) and a second chromosomal translocation involving an insertion (ins) of chromosome 22 material into chromosome 12. Figure 11.1 is a black-and-white

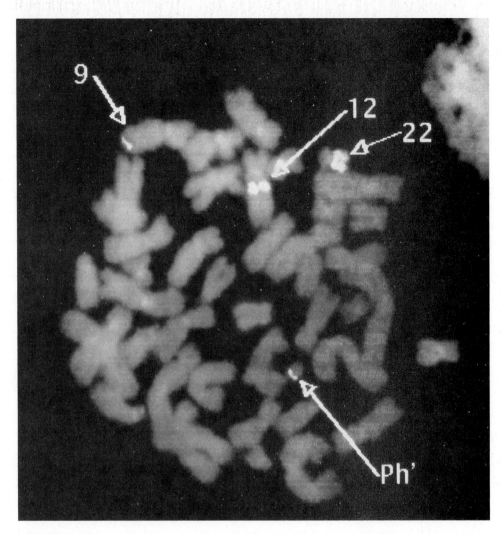

FIGURE 11.1. A black-and-white photograph of the cell shown on the cover but having the chromosome aberrations labeled. (Courtesy of A. Wiktor, J. Zabawski, and D.L. Van Dyke, Department of Medical Genetics, Henry Ford Health System, Detroit, MI.)

photograph of the cell shown on the back cover that has been labeled to reveal these aberrant chromosomes. Figure 11.2 is a karyotype of the same chromosome spread showing the G-banding pattern of the rearranged chromosomes.

Figure 11.3 is a black-and-white schematic representation of a hypothetical karyotype that was constructed from a chromosome spread subjected to FISH using three different DNA probes, each treated so that they will fluoresce a different color. The first DNA probe was a histone H1 gene from *Drosophila*, which fluoresced green in the spread and appears as light gray in the diagram; the second DNA probe was the *Ultrabithorax* gene from *Drosophila*, which fluoresced red in the spread and appears as dark gray in the diagram; and the third DNA probe was the human cystic fibrosis gene (transmembrane conductance regulator), which fluoresced yellow in the spread and appears as white in the diagram. Carefully observe Figure 11.3, paying attention to both the locations of each of the probes and the relative width (thickness) of each band and record your observations in the space provided; then, answer the following questions.

_____

_____

_____

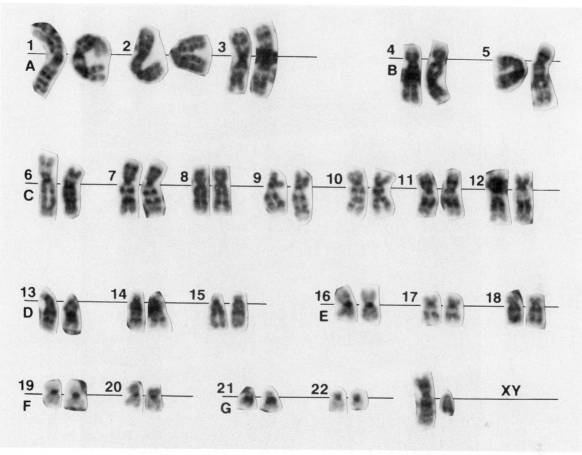

FIGURE 11.2.   A karyotype of G-banded chromosomes of the same cell as the one visualized by FISH and shown in Figure 11.1 and on the cover. (Courtesy of A. Wiktor, J. Zabawski, and D.L. Van Dyke, Department of Medical Genetics, Henry Ford Health System, Detroit, MI.)

1. How many chromosome locations does each gene have in humans and what chromosome(s) are they on?

   a. Histone H1? _____

   b. *Ultrabithorax?* _____

   c. Cystic fibrosis? _____

2. How do you interpret the difference in the width (thickness) of the bands for the different genes and the observation that the histone bands distributed on chromosome 6 are on both sides of that

   centromere? _____

   _____

3. Using Figure 10.1 from Investigation 10, designate the chromosome location(s) for each of these genes.

   a. Histone H1: _____

   b. *Ultrabithorax* (*Hox* genes in humans): _____

   c. Cystic fibrosis: _____

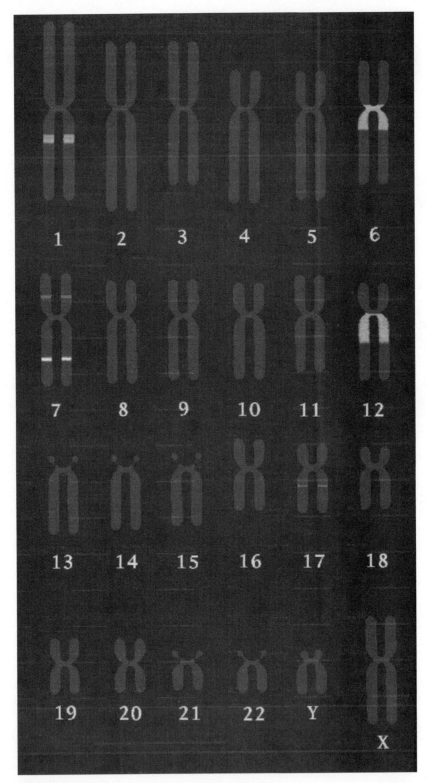

FIGURE 11.3. Diagram of human chromosomes showing *in situ* hybridization sites. See text for details.

C. Physical Maps for the Human Genome

With the development of the Human Genome Initiative in the mid-1980s (see also Investigation 15) and the coordination the initiative provided, it became possible to construct a variety of physical and genetic maps of humans. These maps, although different in the way in which they are constructed and different in the types of information they provide, share the common goal of developing a complete description of the human genome. Although beyond the scope of this laboratory manual, a brief listing and description of these various types of maps is provided. Students are referred to any of a variety of textbooks for readings on the Human Genome Project; for example, Strachan and Read (2004) provide an excellent chapter on this topic.

Different Types of Human Physical Maps[2]

1. **Cytogenetic maps** are chromosome maps based on the various chromosomal banding techniques.
2. **Breakpoint maps** are chromosome maps based on chromosome fragments generated in somatic cell hybrid lines by translocations and deletions.
3. **Restriction maps** are maps of restriction endonuclease sites on various chromosomes.
4. **Cloned contig maps** are maps of overlapping contiguous cloned DNA sequences.
5. **Sequence-tagged maps** are constructed by locating various known sequences on chromosomes.
6. **Expression maps** are maps of expressed human genes as determined by cDNA and *in situ* hybridization.
7. **DNA sequence maps** would represent the final goal of a complete sequence of the base pairs of the human genome.

REFERENCES

ASHBURNER, M., K.G. Golic, and R.S. Hawley. 2005. *Drosophila*: A *laboratory handbook*, 2nd ed. Cold Spring Harbor, NY: Cold Spring Harbor Laboratory Press.

BATESON, W., E.R. SAUNDERS, AND R.G. PUNNETT. 1905. Experimental studies in the physiology of heredity. *Reports to the Evolution Committee of the Royal Society*, II. [Reprinted in Peters, J.A., ed. 1959. *Classical papers in genetics*. Englewood Cliffs, NJ: Prentice-Hall].

CANTOR, C.R., AND C.L SMITH. 1999. *Genomics: The science and technology behind the human genome project*. New York: John Wiley and Sons.

DEMEREC, M., AND B.P. KAUFMANN. 1986. *Drosophila guide*. 9th ed. Washington, DC: Carnegie Institution of Washington.

KLUG, W.S., AND M.R. CUMMINGS. 2006. *Concepts of genetics*. 8th ed. Upper Saddle River, NJ: Prentice-Hall.

McCLATCHEY, K.D., ed. 2002. *Clinical laboratory medicine*, 2nd ed. Baltimore: Williams & Wilkins Co.

McCONKEY, E.H. 2004. *How the human genome works*. New York: Jones and Bartlett Publishers.

McKUSICK, V.A. 1971. The mapping of human chromosomes. *Scientific American* 224(4):104–113.

———. 1980. The anatomy of the human genome, *Journal of Heredity* 71(6):370–391.

———, et al. 1998. *Mendelian inheritance in man: A catalog of human genes and genetic disorders*, 12th ed. Baltimore: Johns Hopkins University Press.

McKUSICK, V.A., and F.H. Ruddle. 1977. The status of the gene map of the human chromosomes. *Science* 196(4288):390–405.

MERTENS.,T.R. 1972. Investigations of three-point linkage. *The American Biology Teacher* 34(9):523–526.

MERTENS, T.R., and A.B. Burdick, 1954. The morphology, anatomy, and genetics of a stem fasciation in *Lycopersicon esculentum. American Journal of Botany* 41:726–732.

MOLNAR, S. 2001. *Human variation, races, types, and ethnic groups*, 5th ed. Upper Saddle River, NJ: Prentice-Hall.

---

[2] Based on Strachan and Read (2004).

Online Mendelian Inheritance in Man, OMIM™ McKusick-Nathans Institute for Genetic Medicine, Johns Hopkins University (Baltimore, MD) and National Center for Biotechnology Information, National Library of Medicine (Bethesda, MD). 2000. http//www.ncbi.nih.gov/omim/

Ruddle, F.H., and R.S. Kucherlapati. 1974. Hybrid cells and human genes. *Scientific American* 231(1):36–44.

Strachan, T., and A.P. Read. 2004. *Human molecular genetics,* 3rd ed. New York: Wiley-Liss.

Sullivan, W., M. Ashburner, and R.S. Hawley. 2000. *Drosophila protocols.* Cold Spring Harbor, NY: Cold Spring Harbor Laboratory Press.

Sutton. W.S., 1903. The chromosomes in heredity. *Biological Bulletin* 4:231–251. [Reprinted in Peters, J.A., ed., 1959. *Classical papers in genetics.* Englewood Cliffs, NJ: Prentice-Hall].

INVESTIGATION 12

Genetics of Ascospore Color in *Sordaria*: An Investigation of Linkage and Crossing Over Using Tetrad Analysis

Tetrad analysis of meiosis and recombination is a procedure that allows an allele marker located on each chromatid of a synapsed tetrad to be followed through the entire meiotic process, including the events of recombination and crossing over if they occur. Two major types of important information can be obtained in this way. The first (using **ordered tetrad analysis**) allows identification of which two of the four chromatids of a tetrad actually participate in crossing over during recombination. The second allows the mapping of a gene marker relative to the centromere of the chromosome, thus positioning a gene on a chromosome relative to a directly observable cytological marker; i.e., the centromere. Once one gene is mapped to the centromere, all other genes showing linkage to that gene can also be assigned a relative position on the chromosome with respect to the centromere.

Sordaria (as with *Neurospora*) is an **ascomycete** fungus that is convenient to use in the classroom laboratory for conducting tetrad analysis of the segregation of ascosore color mutants. *Sordaria* spends most of its life cycle in a haploid vegetative state. However, under certain conditions, two haploid strains can be induced to undergo a sexual process in which they fuse to form diploid zygotes. Instead of dividing mitotically, these diploid zygotes almost immediately undergo meiosis to form haploid **ascospores** that are maintained in a linear order in spore sacs called **asci** (sing., **ascus**). Thus, the order of the chromatids within a tetrad of a diploid zygote is directly reflected in the linear order of the ascospores in an ascus.

In the 1950s, Lindsay Olive, then at Columbia University, initiated studies of ascospore color mutants in *Sordaria*. Dr. Olive induced mutations in the mold using ultraviolet light. He found that ascospore color is autonomously determined by the genotype of the spore itself. Thus, segregation of alleles affecting spore color can be observed directly in the ascus; each haploid spore's phenotype is determined by the spore-color allele that it possesses. Because of the easily recognized phenotype of spore mutants and the ordered behavior of their chromosomes during meiosis, both *Sordaria* and *Neurospora* have been very useful in studying linkage and the mechanisms of recombination.

OBJECTIVES OF THE INVESTIGATION

Upon completion of this investigation, you should be able to

1. **outline** the procedure for mating genetically different strains of *Sordaria*,

2. **calculate** the map distance between a gene and the centromere, given the frequency of the different types of asci resulting from the cross of two genetically different strains of *Sordaria*, and

3. **discuss** gene conversion and its implications for mechanisms of recombination.

Materials needed for this investigation:

For each student:

compound microscope

microscope slides

cover glasses

dissecting needles

petri dish containing crossed strains of *Sordaria*

small bottle of 70% ethyl alcohol

alcohol lamp or bunsen burner

For the class in general:

wild-type *Sordaria fimicola* having black ascospores

mutant *S. fimicola* having gray ascospores

media for culturing and crossing *Sordaria* (see *Instructor's Manual*)

I.   PREPARATION

Approximately 7–12 days before today's laboratory period, a cross was made between the wild-type and gray-ascospore strains of *Sordaria*. The following procedure was used in making the cross: A wax pencil was used to mark into fourths the bottom of the petri dish containing sterile medium. The quadrants so created were alternately labeled ╪ and *g*, and inoculations were made accordingly, with two different strains always adjacent to each other. A narrow, stainless steel spatula was flamed, cooled in 70% ethyl alcohol, and again flamed slightly to remove the excess alcohol. The sterile spatula was used to cut from the stock culture of the wild-type strain a small piece of agar (approximately 3–5 mm square) containing some of the *Sordaria* mycelium. This piece of agar was transferred to the center of a section of the petri dish marked ╪. After both quadrants of the petri dish labeled + had been inoculated, the procedure was repeated with the mutant strain. The petri dish was then placed in a darkened incubator at 25°C.

Within 7–12 days, the petri dish appears as in Figure 12.1. The two strains of *Sordaria* have grown over the surface of the agar. Sexual reproduction occurs where they meet. The products of this sexual mating are black fruiting bodies called **perithecia** (sing., **perithecium**) (Figure 12.2). The fact that sexual reproduction has occurred can also be seen in Figure, 12.1; the black dots in the illustration are the perithecia.

FIGURE 12.1. Petri dish showing the crossed strains, ╪ and *g*, of *Sordaria fimicola* and the line of hybrid perithecia, represented by the black dots on the mycelium.

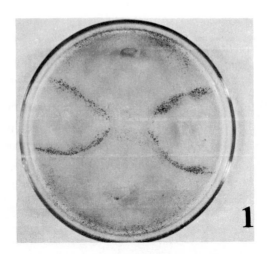

FIGURE 12.2. Wet mount of a mature perithecium in the center and a small, immature perithecium to the left (200×).

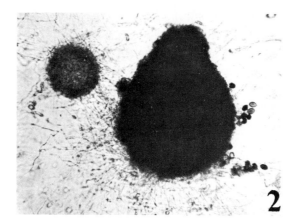

II. PROCEDURE

Using sterile technique (i.e., flaming a teasing needle, cooling in alcohol, reflaming), remove a perithecium from the petri dish and transfer it to a drop of water on a clean microscope slide. Put a cover glass on the drop of water. Now, gently apply pressure to the cover glass with the eraser end of a pencil to squash the perithecium and eject its contents.

Next, examine the slide under low power and then under high power. What you observe should resemble Figure 12.3. Ejected from the crushed perithecium are many slender, saclike asci. Each ascus contains all of the meiotic products or ascospores produced when one diploid zygote underwent meiosis. *Note:* The zygote is the only diploid cell in the life cycle of this organism.

FIGURE 12.3. Wet mount showing the various ascospore arrangements in the asci that have been ejected from the perithecia (200×).

III. COLLECTING DATA

You will note that the arrangement of gray and black ascospores differs in the various asci.

1. How many different arrangements do you find? _____
   What are they? (In answering this question, list the black spores as *+* and the mutant spores as *g*;

   e.g., *++++gggg* is a common ascospore arrangement.) _____

   _____

   _____

Ascospore patterns. The production of an ascus containing eight ascospores from a diploid zygote involves four nuclear divisions. The first two divisions are meiotic and produce four haploid nuclei. The third division is mitotic and results in the four haploid nuclei giving rise to eight ascospore nuclei. The fourth division is also mitotic and occurs after ascospore delimitation so that the mature spore is binucleate (Carr and Olive, 1958).

The first three nuclear divisions that produce the ascus and the various possible ascospore patterns of hybrid asci are depicted in Figure 12.4. When no crossing over occurs between the gene and the centromere, segregation of the alleles at the first meiotic division (i.e., when homologous centromeres segregate at anaphase I—first-division segregation) results in the sequence $4+:4g$ or $4g:4+$. When crossing over occurs, the other ascospore arrangements follow second-division segregation (i.e., when the centromere duplicates at anaphase II and the sister chromatids separate). The $2+:2g:2+:2g$ arrangement is the result of crossing over between chromatids two and three (Figure 12.4), whereas if chromatids one and four cross over, a $2g:2+:2g:2+$ sequence results. The pattern of $2+:4g:2+$ results from crossing over involving strands two and four, whereas crossing over between chromatids one and three gives the pattern $2g:4+:2g$.

Now examine your slide carefully at 430× and count the number of asci having each of the arrangements that we have discussed and that are illustrated in Figure 12.4. The instructor will place a table on the chalkboard in which to record your data as part of the cumulative class data. Record your data and those accumulated for the entire class in Table 12.1.

Map distance. The percentage of **crossing over (map distance)** between the gene, *g*, and the centromere is estimated on the basis of the frequency of second-division segregation. The percentage of second-division segregation can be calculated from the data in Table 12.1 in the following manner: Determine the total number of second-division segregation asci and divide this number by the total number of asci counted. Multiply the resulting number by 100 to yield the percentage of second-division segregation.

TABLE 12.1. Ascospore Distribution Patterns in *Sordaria fimicola* at 25°C

First-division Segregation	Your Data	Class Data
$4+:4g$	_____	_____
Second-division segregation	Your Data	Class Data
$2+:2g:2+:2g$	_____	_____
$2+:4g:2+$	_____	_____
$2g:4+:2g$	_____	_____
Totals	_____	_____

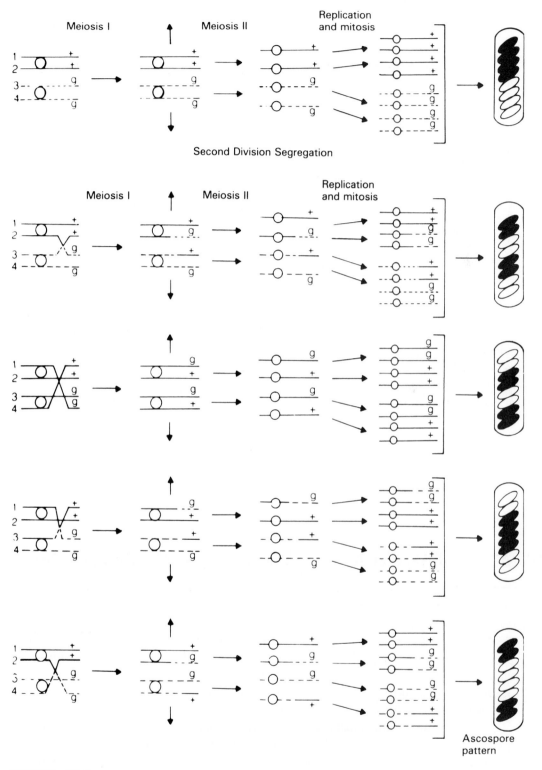

FIGURE 12.4. Illustration of the first three nuclear divisions that produce various possible arrangements of ascospores in the ascus after first- and second-division segregation. See text for detailed explanation. *Note:* Each line in the figure represents a chromosome or chromatid. Circles represent centromeres.

2. Now make these calculations for your own and for the class data.

	Total number of second-division segregation asci	Total number of asci counted	Percentage of second-division segregation
Your data	_____	_____	_____
Class data	_____	_____	_____

Recall (from Figure 12.4) that second-division segregation asci result from a single crossover between the *g* locus and the centromere. Because only two of the four chromatids cross over, **one-half** of the chromatids are therefore **recombinant** and one-half are **parental**. The measure of the distance between the gene locus and the centromere (i.e., the map distance) is thus **one-half** the percentage of second-division segregation. On the basis of this information and the data obtained in this investigation, what is the map distance between the *g* locus and the centromere for your data?

_____

3. For the class data? _____

4. Which allele is dominant, the + allele or *g*? Explain.

_____

5. Did you find any perithecia in which the asci contained eight black or eight gray ascospores?

How could these spore arrangements be explained? _____

_____

_____

_____

IV. ATYPICAL SEGREGATION RATIOS AND GENE CONVERSION

With Section III we observed how meiosis in *Sordaria* results in a 4:4 ascospore ratio when no recombination has occurred, whereas recombination results in various 2:2:2:2 or 2:4:2 patterns. Occasionally, however, aberrant tetrad ratios occur. These include 3:1:1:3, 5:3, and 6:2 (all aberrant 4:4) ascospore patterns. Examine Figure 12.3 to see if you can find any of these aberrant arrangements. Only an alteration of genetic material during meiosis, such as a change of wild type to mutant or mutant to wild type, can account for these ascospore patterns. The frequency of these changes is too great, however, to be explained by spontaneous mutation. Consequently, the term **gene conversion** has been applied to these ascospore alterations.

You previously learned that the eight ascospores of an ascus are derived from mitosis of the four products of the second meiosis. Thus, each ascospore represents one-half of the DNA double helix present in the second meiotic division chromatids. Asci possessing 3:1:1:3 or 5:3 ascospore patterns, therefore, have two ascospores derived from a chromatid that possessed a "hybrid" DNA molecule. (The strands of the DNA double helix have a noncomplementary nucleotide sequence for the gene in question.) The 6:2 ascospore pattern results from a complete chromatid change.

Aberrant ascospore patterns and the process of gene conversion have provided important information on possible molecular mechanisms of recombination. One possible mechanism of recombination that could account for gene conversion is based on the Holliday model and is shown in Figure 12.5.

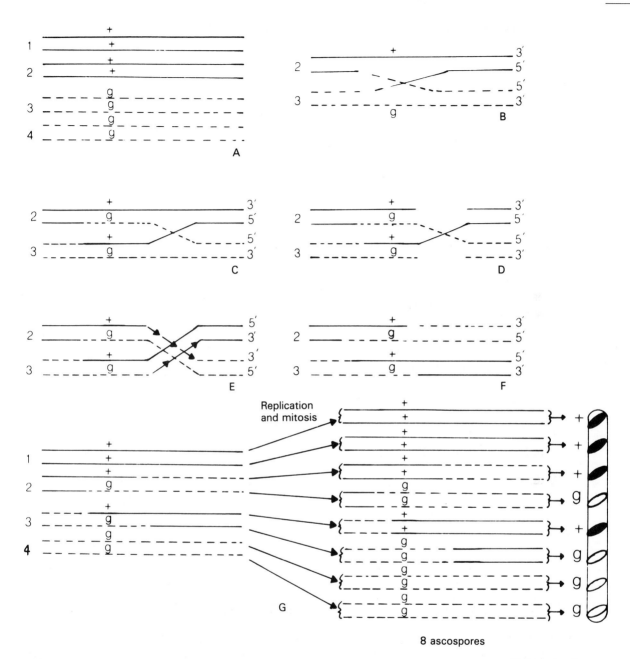

FIGURE 12.5.   Diagram showing one possible molecular mechanism for gene conversion, based on Holliday's model for recombination. (A) DNA double helices of four chromatids of a tetrad. (B) Only chromatids 2 and 3 of the tetrad are shown, with the 5′ and 3′ ends of the DNA strands indicated. An endonuclease nicks one strand of each double helix, and the strands partially separate. (C) Nicked single-stranded DNA molecules then base pair with the opposite chromatid DNA molecules. (D) Exonuclease removes unpaired nucleotides of the nonrecombinant strands. (E) DNA polymerases synthesize new DNA to fill the gaps created by the exonuclease. DNA molecules showing this configuration are called *DNA heteroduplexes.* (Diagram shows a twisting of the DNA molecule in each chromatid.) (F) Recombinant chromatids separate to form hybrid DNA molecules for each chromatid (see areas labeled + and *g*). *Note:* If any mismatching of bases occurs, DNA repair mechanisms may correct one or both DNA strands. (G) Diagram showing the DNA molecules of the four second meiotic chromatids (at left) and the eight ascospores produced by the postmeiotic division (at right).

1. Did you detect any ascospore arrangements showing 3:1:1:3 (aberrant 4:4) ratios? If so, what do these represent? With what frequency did you encounter this arrangement? _____

   _____

   _____

   _____

2. By following the incorporation of tritiated thymidine, researchers have demonstrated that a small amount of DNA synthesis occurs during pachynema. What function might this DNA synthesis serve?

   _____

3. Did you detect any 5:3 or 6:2 arrangements? _____

   a. How can the 5:3 or 3:5 arrangement be explained? _____

      _____

   b. What might account for the 6:2 or 2:6 patterns? _____

      _____

4. Did you detect any ascospore arrangements other than the ones previously mentioned? If so, how might these be explained? _____

   _____

   _____

5. What was the total frequency of aberrant asci in your data sample?

   _____

   In the total class sample? _____

REFERENCES

BINKLEY, S.W. 1978. Genetic recombination in *Sordaria*. *Carolina Tips* 41(8):29–30.

CARR, A.J.H., and L.S. OLIVE. 1958. Genetics of *Sordaria fimicola*. II. Cytology. *American Journal of Botany* 45:142–150.

CASSELL, P., and T.R. MERTENS. 1968. A laboratory exercise on the genetics of ascospore color in *Sordaria fimicola*. *The American Biology Teacher* 30(5):367–372.

FINCHAM, J.R.S. 1983. *Genetic recombination in fungi*, 2nd ed. Carolina Biology Reader Series. Burlington, NC: Carolina Biological Supply Co.

HOLLIDAY, R. 1964. A mechanism for gene conversion in fungi. *Genetical Research* 5:282–304.

KLUG, W.S., and M.R. CUMMINGS. 2006. *Concepts of genetics*, 8th ed. Upper Saddle River, NJ: Prentice Hall.

OLIVE, L.S. 1956. Genetics of *Sordaria fimicola*. I. Ascospore color mutants. *American Journal of Botany* 43:97–107.

RUSSELL, P.J. 2006. *iGenetics: a molecular approach*, 2nd ed. New York: Benjamin Cummings.

SUZUKI, D.T., et al. 2004. *An introduction to genetic analysis*, 8th ed. New York: W.H. Freeman.

WALDMAN, A.S. 2004. *Genetic recombination: reviews and protocols*. Totowa, NJ: Humana Press.

INVESTIGATION 13

Open-Ended Experiments Using *Drosophila*: Locating a Mutant Gene in Its Chromosome

In Investigations 1 and 8, *Drosophila* was used to demonstrate basic principles related to simple Mendelian genetics. Mendel's laws of segregation and independent assortment were studied in monohybrid and dihybrid crosses, and the extension of Mendelism to sex linkage was investigated. In the present investigation, *Drosophila* will be used in a more experimental and open-ended fashion, rather than merely to demonstrate established principles.

To accomplish this goal of open-endedness, the instructor will code mutant names so that you will not know what mutants are being used. Each mutant will then be essentially an unknown, and you will be required to approach the laboratory work as an experiment or investigation and not merely as an exercise for which the conclusions are known before the exercise is initiated.

Also, in this investigation the instructor will select crosses that are sufficiently sophisticated to challenge the advanced student to think critically in analyzing experimental data. Typically, these crosses will yield results that you will not readily anticipate. Unusual ratios may be obtained, and the ratios for males may differ from those for females. The crosses may involve epistasis (gene interaction), linkage, and either autosomal or X-linked genes. A combination of these variables can effectively create the illusion of an original investigation. Certainly for you, as students, the investigation will be open-ended and you may need to pursue a variety of hypotheses to explain your data.

The general goal of these open-ended experiments is for you to determine the mechanism of inheritance of the mutant traits involved in the cross. The final product of your work will be a report consisting of the parts usually found in a scientific publication: introduction, literature review, materials and methods, experimental data, discussion, summary, conclusions, and literature cited.

OBJECTIVES OF THE INVESTIGATION

Upon completion of this investigation, you should be able to

1. **analyze** F_1 and F_2 data resulting from relatively sophisticated *Drosophila* experiments in which the genetic traits are given to you as unknowns,

2. **write** a scientific paper following a suggested format to summarize a *Drosophila* experiment, and

3. **outline** an experimental protocol that can be used to assign a specific mutant gene to a specific chromosome in *Drosophila*.

Materials needed for this investigation:

For each student:

materials for handling *Drosophila* (see Investigation 1)

For the class in general:

selected *Drosophila* stocks (to be chosen by the instructor); record their phenotypes and stock numbers in the following space._____

_____

_____

Stock of *Cy/Pm* ; *D/Sb* flies

Culture bottles containing medium

I.   CROSS INVOLVING UNKNOWNS

Your instructor will select stocks appropriate to this part of the laboratory work. This manual will not suggest any particular experiments; students will use different crosses on different occasions. *References given at the end of this investigation will help provide suggestions for appropriate experiments.* Record the data for this experiment in Table 13.1.

Upon completion of this investigation, report your results in a scientific paper consisting of the following parts.

1. **Introduction.** State the purpose of the experiment in one sentence and indicate the experiment that was performed.

2. **Literature review.** Summarize any literature—books or periodicals—that you have read that has a direct bearing on the particular experiment you are reporting. Cite the literature to which you refer, using the author-date method employed in this manual. List all literature cited at the end of your paper.

3. **Materials and methods.** Briefly describe the stocks of flies with which you worked and outline the procedures you used to do the experiment.

4. **Experimental data.** Include in this section all of the important observations pertaining to this experiment. Be certain to give the following information: phenotypes of the parent flies and the date on which the cross was made; phenotypes of the F_1 flies; numbers of flies in each phenotypic class in the F_2 generation. Include Table 13.1 as a means of reporting pertinent data relative to this experiment.

5. **Discussion.** In the discussion, state your hypothesis and outline the theoretical expectation for the experiment. Do this by diagramming the cross, showing genes on chromosomes. Answer the following question in the discussion: Do the experimental data fit the theoretical expectation? To answer this question, it will be necessary to perform a statistical analysis of the data using the chi-square test. If the observed data do not fit the theoretical expectation, try to arrive at a reasonable explanation (one based on Mendelian principles) for this variation. You may wish to review appropriate literature to help you write a good discussion.

6. **Summary and conclusions.** Briefly summarize the experiment in one paragraph, and then number and list the significant conclusions that can be drawn from the data. These conclusions should pertain to the mechanism of inheritance of the traits studied in the experiment.

7. **Literature cited.** This section of the report should provide complete bibliographic entries for all literature cited in the Literature Review and Discussion. Give author or authors, date of publication, title of article, journal name, volume, and pages. When citing a book, include author, date, title, the name of the publisher, and the place of publication.

TABLE 13.1. Record of a *Drosophila* Experiment

Record data obtained in experiment involving unknown mutants. Be certain to record both phenotypes and stock numbers of parent (P_1) flies.

Experiment number _____   **Name** _____

1. Cross _____ female $\times$ _____ male

2. Date P_1s mated: _____

3. Date P_1s removed: _____

4. Date F_1s first appeared: _____

5. Phenotype of F_1 males: _____

6. Phenotype of F_1 females: _____

7. Date F_1 male and female placed in fresh bottle (or date F_1 virgin female testcrossed): _____

8. Date F_1 flies removed: _____

9. Date F_2 (or testcross) progeny appeared: _____

10. Record F_2 or testcross data in the following table:

	Males		Females		
	Phenotype	Number	Phenotype	Number	Total
a.	_____	_____	_____	_____	_____
b.	_____	_____	_____	_____	_____
c.	_____	_____	_____	_____	_____
d.	_____	_____	_____	_____	_____
e.	_____	_____	_____	_____	_____
f.	_____	_____	_____	_____	_____
g.	_____	_____	_____	_____	_____
h.	_____	_____	_____	_____	_____
Totals		_____		_____	_____

II. LOCATING AN UNKNOWN *DROSOPHILA* MUTANT IN A PARTICULAR CHROMOSOME

The purpose of this experiment is to determine in which of the four linkage groups (chromosomes) of *Drosophila melanogaster* a particular mutant gene is located. For this purpose you will use a special stock of flies designated as *Curly/Plum, Dichaete/Stubble*.[1] This stock is perpetually heterozygous for the four dominant mutant genes *Cy, Pm, D*, and *Sb*, each of which is lethal when homozygous. The genotype of these flies is

$$\frac{Cy+}{+Pm} \quad \frac{D+}{+Sb}$$

Cy (curly wing) and *Pm* (plum eye color) are located on the second chromosome, whereas *D* (dichaete wing) and *Sb* (stubble bristles) are third-chromosome mutants.

No crossing over will occur between *Cy* and *Pm* (nor between *D* and *Sb*), because these genes are associated with chromosome inversions that in effect prevent crossing over.[2] *Cy/Pm D/Sb* is a balanced lethal stock; the flies breed true when crossed *inter se*.

1. What gametes will a *Cy/Pm D/Sb* fly produce? Write the genotypes of these gametes.

   _____

   _____

   _____

   _____

   _____

2. In the space that follows make a Punnett square (checkerboard) to show that *Cy/Pm D/Sb* flies will breed true when crossed *inter se*.

3. **Initial cross**: Now become familiar with the phenotype of *Cy/Pm D/Sb* flies and with the phenotype of the mutant stock of flies your instructor has chosen for you to identify as to linkage group.

   To locate the selected mutant gene in its linkage group or chromosome, proceed as follows: Select virgin females homozygous for the mutant gene (arbitrarily designate the allele as *m*) that you are to locate. Mate such virgin mutant females to *Cy/Pm D/Sb* males.

---

[1] The *Curly/Plum, Dichaete/Stubble* stock of *D. melanogaster* is available from Carolina Biological Supply Co., 2700 York Road, Burlington, NC 27215.

[2] Actually, inversions do not prevent crossing over. Crossing over occurs, but the meiotic products are nonviable because of chromosomal deletions, and duplications, dicentric bridges, and acentric fragments.

4. What phenotypes can be expected in the F_1 generation, assuming the mutant gene (m) to be an autosomal recessive gene? _____

_____

_____

a. Will different F_1 results be obtained if m is X-linked rather than autosomal? If so, describe.

_____

_____

b. What will be the F_1 phenotypes if the mutant gene is a dominant?

_____

5. What F_1 phenotypes were actually observed? _____

_____

6. **Testcross:** As you noted in answering question 5, you obtained several F_1 phenotypes. You are now going to select F_1 males of one of these phenotypes to mate with virgin female flies homozygous for the mutant trait; that is, you are going to perform a testcross. In selecting the F_1 males for the testcross, use those males having phenotypic characteristics that are least likely to be confused with the mutant trait. For example, if m is an eye color mutant, it would not be wise to select the *Plum*-eyed F_1 males. Or, if the mutant affects the wing size or shape, it would not be wise to select F_1 flies having *Curly* wings.

7. Now, indicate what testcross you have actually performed, giving the phenotype of the F_1 males used. _____

_____

8. What phenotypes did you observe among the testcross progeny? _____

_____

What testcross ratio did you observe? _____

9. How can the testcross results be used to determine the chromosome on which the mutant m is located? To answer this question, consider the following specific questions:

a. What testcross results would you expect if the mutant gene is X-linked? (Consider your response in 4a above.) _____

_____

b. What testcross results would be obtained if the mutant m is located on the same chromosome (i.e., chromosome II) as *Cy* and *Pm*? _____

_____

c. What testcross results would be obtained if the mutant m is located on the same chromosome as *D* and *Sb* (i.e., chromosome III)? _____

_____

d. If the mutant *m* is not located on the X chromosome, nor on chromosome II nor on III, but is on chromosome IV, what testcross results may be expected? _____

_____

10. Finally, on the basis of your testcross data and your answers to questions 9a through 9d, indicate on what chromosome the mutant gene *m* is actually located. _____

_____

_____

11. Write a summary statement in which you explain how the *Cy/Pm D/Sb* stock is used to locate a mutant gene in a particular chromosome.

_____

_____

_____

_____

_____

REFERENCES

ASHBURNER, M., K.G. GOLIC, and R.S. HAWLEY 2005. *Drosophila: a laboratory handbook*, 2nd ed. Cold Spring Harbor, NY: Cold Spring Harbor Laboratory Press.

DEMEREC, M., and B.P. KAUFMANN. 1986. *Drosophila guide*, 9th ed. Washington, DC: Carnegie Institution of Washington.

GRAF, U., and N. VAN SCHAIK. 1992. *Drosophila genetics, A practical course*. New York: Springer-Verlag.

KLUG, W.S., and M.R. CUMMINGS. 2006. *Concepts of genetics*, 8th ed. Upper Saddle River, NJ: Prentice Hall.

KOHLER R.E. 1994 *Lords of the fly: Drosophila genetics and the experimental life*. Chicago, IL: The University of Chicago Press.

MERTENS, T.R. 1969. Open-ended genetics laboratory investigations using *Drosophila melanogaster*. *The Science Teacher* 36(9):68–70.

———. 1976. An open-ended *Drosophila* genetics experiment. *The American Biology Teacher* 38(5):288–291.

———. 1980. Investigating a new *Drosophila* stock. *Carolina Tips* 43(1):1–2.

———. 1983. Open-ended laboratory investigations with *Drosophila*. *The American Biology Teacher* 45(5):264–266.

———. 1987. The inheritance of white eye color in *Drosophila. Carolina Tips* 50(8): 29–30.

PECHENIK J.A. 2004. *A short guide to writing about biology*, 5th ed. New York: Longman/Pearson.

ROBERTS D., and D.B. ROBERTS, eds. 1998. *Drosophila: a practical approach (Practical approach series)*, 2nd ed. New York: Oxford University Press.

STRICKBERGER, M.W. 1962. *Experiments in genetics with Drosophila*. New York: John Wiley & Sons.

SULLIVAN W., M. ASHBURNER and R.S. HAWLEY. 2000. *Drosophila protocols*. Cold Spring Harbor, New York: Cold Spring Harbor Laboratory Press.

INVESTIGATION 14

The Genetic Material: Isolation of DNA

In 1944, Avery, MacLeod, and McCarty demonstrated that deoxyribonucleic acid is the genetic material of the pneumococcus bacterium. Although this finding was not fully appreciated at the time, the next decade saw the realization of the significance of DNA as the genetic chemical and the development by James Watson, Francis Crick, and Maurice Wilkins of a model of the structure of the DNA molecule. Inherent in the Watson-Crick model was a proposal for how DNA could replicate, how it could encode information, and how changes in its structure—mutations—could produce phenotypic changes in organisms.

Most undergraduates studying biology over the past 50 years have learned a great deal about DNA and the so-called generalized central dogma of molecular genetics: DNA → RNA → protein → trait. In addition to replication, DNA serves as a template for the synthesis of three major types of RNA. Transfer RNA (tRNA) molecules are small molecules (75–80 nucleotides) that bind to particular amino acids and that recognize corresponding triplets (codons) in mRNA. Messenger RNA (mRNA) contains sequences of three nucleotides (triplets) that specify different amino acids in a protein. Different types of ribosomal RNA (rRNA) together with proteins make up the bulk of each ribosome, an ultramicroscopic organelle that is the assembly site where amino acids are bonded together to form a polypeptide.

What is perhaps surprising is that, in learning a great deal about DNA, RNA, and protein synthesis, many biology students have never actually seen high-molecular-weight DNA nor directly observed any of its properties. This investigation will provide you with the opportunity to isolate and observe DNA and to determine its concentration in a solution derived from beef spleen or liver.

OBJECTIVES OF THE INVESTIGATION

Upon completion of this investigation, you should be able to

1. **extract** DNA from tissue prepared from a beef spleen or liver,

2. **describe** the appearance of the DNA so extracted,

3. **determine** the milligrams of DNA extracted per gram of tissue used initially, and

4. **operate a spectrophotometer** to obtain the data needed to achieve objective 3.

Materials needed for this investigation:

For each group of four students:

8 test tubes

1 stopper for a test tube

7 aluminum test tube caps or aluminum foil

1 250-ml beaker

1 1000-ml beaker

1 glass stirring rod

1 50-ml plastic screwcap centrifuge tube

water bath at 100°C (boiling water)

Spectronic 20 or other spectrophotometer

spectrophotometer tubes

For the class in general:

ice and ice water

supply of pasteur pipettes and bulbs

standard stock solution of DNA containing 1 mg DNA/ml (prepared from commercial calf thymus DNA)

supply of 5-ml pipettes

1 automatic pipetter of Dische reagent

1 centrifuge equipped with a rotor to handle 50-ml tubes

cheesecloth and 1000-ml beaker

1 blender

1000 ml cold saline citrate solution:

   2.94 g Na citrate

   8.12 g NaCl

   dissolve in 1000 ml distilled water and pH to 7.0–7.4 with HCl or NaOH

1000 ml cold 2.6 M NaCl solution:

   152.1 g in 1000 ml distilled water

200 ml Dische reagent:

   2.0 g diphenylamine

   200 ml glacial acetic acid

   then add 5.5 ml conc. H_2SO_4

1000 ml freezer-cold 95% ethanol

I. USING A SPECTROPHOTOMETER TO MEASURE OPTICAL DENSITY

Optical density (absorbance) refers to the ability of a solution to prevent light of a certain wavelength from passing through it. You will measure the concentration of DNA isolated from the beef spleen solution by measuring the optical density of that DNA after it has reacted with Dische reagent. The darker the stain reaction (the higher the concentration of DNA), the greater is the absorbance. The spectrophotometer should be set at 500 nm,[1] the optimal wavelength for the color of the solution in this experiment. According to **Beer's law**, there is a direct relationship between the concentration of a solution and absorbance. This relationship can be expressed as $A = kc$, where A = absorbance, k = constant of the instrument, and c = concentration.

Before this experiment begins, become familiar with the Spectronic 20 spectrophotometer by studying the following directions and referring to Figure 14.1. If another model of spectrophotometer is to be used, your instructor will provide you with directions for its operation.

Directions for Using the Spectronic 20 Spectrophotometer (Refer to Figure 14.1 to identify the following parts of the spectrophotometer.)

  1. **Power**. Rotate the light source switch clockwise (pilot light will glow) and allow the instrument 5 minutes warmup time.

---

[1] nm = nanometer = one-billionth of a meter.

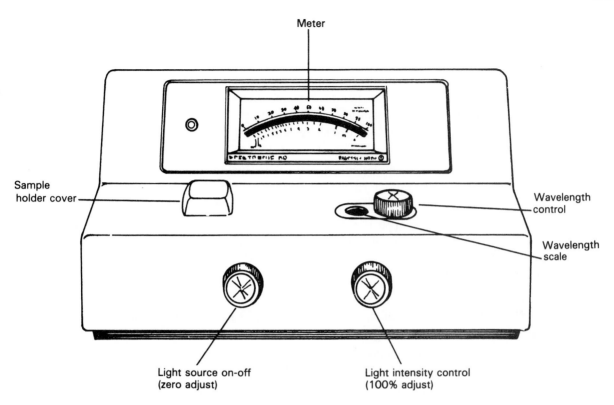

FIGURE 14.1. Spectronic 20 spectrophotometer.

2. **Wavelength selection**. Adjust wavelength control to the desired wavelength setting as indicated on the scale. *Note*: Some spectrophotometers have a filter selection lever located on the lower left corner, which should be set for the wavelength range that corresponds to 500 nm.

3. **Zero setting**. Adjust the zero (light source) control as required until the meter reads 0% transmittance. *Note:* The instrument must not contain a spectrophotometer tube, and the cover of the sample holder must be closed.

4. **Standardizing light control**
   a. Fill a spectrophotometer tube about half way full of the blank solution from tube 5 (see item 16j in Section II below). **Wipe fingerprints from tube**.
   b. Insert the tube into the sample holder, aligning the mark on the tube with the line on the sample holder. Close the cover.
   c. Adjust the light control as required until the meter reads 100% transmittance.

5. **Sample measurement**
   a. Fill a second spectrophotometer tube about half-way full of the sample to be measured.
   b. Remove the tube of the blank solution and replace it with the tube containing the sample. Align the tube as described in step 4b. Close the cover. The ABSORBANCE may now be directly read from the meter. *Important*: When operating at a fixed wavelength, use the blank solution as described in step 4 to check periodically for meter drift from 100% transmittance.

II. EXTRACTION AND QUANTIFICATION OF DNA

Beef spleen or liver from freshly slaughtered animals can be obtained from local slaughterhouses, returned to the laboratory in ice, and used to prepare a DNA solution as outlined in the following steps. Spleen or liver are used because tissue of these organs contains a high percentage of nuclear material (and thus of DNA). Use the following instructions to extract and quantify the DNA. The *Instructor's Manual* provides additional information.

1. Chop up fresh spleen tissue[2] into small cubes (1 cm[3]). Avoid using any fat or connective tissue, which appears lighter in color. Use 10 g tissue for every 100 ml DNA solution required (1 g per 10 ml). Approximately 50 ml of solution will be required for each group of four students.

2. For 100-ml DNA solution, place 100 ml cold citrate-saline buffer into a chilled blender. Drop pieces of tissue into the blender after it has been turned on. Continue to blend for approximately 2–3 minutes. After homogenizing the spleen, pass the homogenate through several layers of cheesecloth into a beaker to remove any large clumps of tissue.

3. Distribute the homgenate between 50-ml centrifuge tubes (approximately 50 ml per tube), and balance the tubes before centrifuging. Mark the fluid level on each tube using a wax pencil. *Note:* If foam is present, wait approximately 1–2 minutes for the foam to rise to the top of the tube, and then remove the foam with a pasteur pipette and add additional homogenate.

4. Centrifuge for 10 minutes at 6000 rpm, preferably in a refrigerated centrifuge. A clinical centrifuge equipped with a rotor for 50-ml tubes also may be used. If this is the case, centrifuge at the highest speed setting for 10 minutes.

5. Discard supernatant.

6. Pour cold citrate-saline buffer into each centrifuge tube to the level of the wax mark and resuspend the pellet. Use a glass rod to break up the pellet, cap the tube, and shake vigorously for a few minutes. Balance tubes before centrifuging.

7. Centrifuge for 10 minutes at 6000 rpm.

8. Discard supernatant.[3]

9. Pour 2.6 M NaCl solution into each tube to the level of the wax mark. Use a glass rod and vigorous shaking to dissolve the pellet. Because there is a lot of DNA present, this step may require several minutes to completely dissolve the DNA from the pellet. Balance tubes.

10. Centrifuge for 15 minutes at 6000 rpm.

11. Pour the supernatant into a 250-ml beaker. Make an ice bath using the 1000-ml beaker and place the 250-ml beaker containing the DNA solution into this ice bath. **Keep the DNA solution chilled**.

12. Quickly dump 100 ml freezer-cold 95% ethyl alcohol into the beaker containing the DNA solution so that the alcohol forms a layer on top of the DNA solution. The DNA will precipitate at the interface between the alcohol and the DNA solution. Describe the appearance of the resulting

    DNA precipitate. _____

    _____

13. Using a glass stirring rod, stir the precipitate that forms at the interface of the DNA solution and the alcohol, winding the precipitate onto the stirring rod. This threadlike precipitate is the DNA. Be sure to extract all of the DNA that can be wound onto the glass rod. What physical characteristics can be ascribed to this DNA based on the way you have extracted it? For example, the fact that you can wind the DNA onto a glass rod implies what about its physical nature? _____

    _____

    _____

    _____

    _____

---

[2] Quick-frozen material from a slaughterhouse may be used and will give a good yield of DNA, but tissue purchased from supermarkets will give reduced yields because of DNA degradation.

[3] If time permits, a third washing of the pellet will produce a cleaner DNA. This is accomplished by repeating steps 6, 7, and 8.

14. Remove the stirring rod with the attached DNA from the beaker, and then gently drain and blot the alcohol from the DNA onto a paper towel. Removal of the alcohol will facilitate dissolving the DNA in water.

15. Transfer the DNA to a test tube containing 10 ml distilled water, and gently remove the DNA from the glass rod by agitating the glass rod up and down. Try not to clump the DNA. Stopper the tube and shake vigorously until the DNA is in solution. Prolonged and vigorous shaking will be required to get all of the DNA into solution.

16. Determine the concentration of DNA in this solution, using the following serial dilution and quantification procedures.
    a. Label seven test tubes from 1 through 7.
    b. Place 2 ml distilled water in tubes 2 through 5, and in tube 7, but not 6.
    c. Place 2 ml **standard stock solution** containing 1 mg DNA/ml in tubes 1 and 2. Set aside tube 1. Mix the contents of tube 2.
    d. Transfer 2 ml from tube 2 to tube 3.
    e. Mix the contents of tube 3 and transfer 2 ml from tube 3 to tube 4.
    f. After mixing tube 4, discard 2 ml from tube 4. Note that tubes 1 through 5 all have 2 ml solution containing from 1 mg DNA/ml (tube 1) to no DNA (tube 5).
    g. Place 2 ml of your isolated and resuspended DNA solution from step 15 above into tubes 6 and 7.
    h. Mix the contents of tube 7 and discard 2 ml of solution. (Tube 7 now has one-half the concentration of tube 6.) These two tubes provide two readings of your isolated DNA, one of which should fall within the readable range of the spectrophotometer.
    i. Mix 4 ml Dische reagent with the contents of each of the seven tubes, and cap the tubes with aluminum caps or aluminum foil to prevent evaporation. Now, heat the tubes in boiling water for 10 minutes.
    j. After cooling the tubes in ice water for a few minutes, pour their contents into spectrophotometer tubes bearing corresponding numbers. You are now ready to read absorbances on a spectrophotometer. (In some cases your instructor may have you use fewer spectrophotometer tubes and will explain the procedure.)
    k. Use tube 5 (distilled water + Dische reagent) as a blank to zero the spectrophotometer, which is set at 500 nm (see directions for use of the spectrophotometer).
    l. Read each of the seven tubes at 500 nm, recording absorbances in the following spaces.

Tube number	Absorbance	DNA concentration (mg/ml)
1	_____	_____
2	_____	_____
3	_____	_____
4	_____	_____
5	_____0_____	_____0_____
6	_____	_____
7	_____	_____

    m. Plot the absorbances of tubes 1 through 4 on a photocopy of the standard graph paper provided on page 305. Plot absorbance on the vertical axis and DNA concentration (milligrams per milliliter) on the horizontal axis. Use this standard curve to determine the concentration of the DNA solution prepared from beef spleen by plotting the absorbance for tubes 6 and 7, determining what concentration of DNA will produce that absorbance. Remember, tube 7 has one-half the concentration of your isolated DNA solution, so multiply the determined concentration by a factor of 2.

n. Remembering that each 10 ml of the DNA-rich solution with which you started contained the DNA extracted from 1 g tissue, calculate the concentration of DNA per gram of bovine tissue and the percentage of the total tissue this DNA amounts to. Record your calculations in the following spaces.

_____

_____

_____

_____

REFERENCES

Avery, O.T., C.M. MacLeod, and M. McCarty. 1944. Studies on the chemical nature of the substance inducing transformation in pneumococcal types. *Journal of Experimental Medicine* 79:137–158.

Johnson, S., and T.R. Mertens. 1989. An interview with Nobel laureate Maurice Wilkins. *The American Biology Teacher* 51(3):151–153.

Kirchen, R.V. 1981. DNA—the master molecule. *Carolina Tips* 44(6):25–27.

Klug, W.S., and M.R. Cummings. 2006. *Concepts of genetics*, 8th ed. Upper Saddle River, NJ: Prentice Hall.

Lewin, B. 2004. *Genes VIII.* Upper Saddle River, NJ: Pearson Prentice Hall.

Pierce, B.A. 2005. *Genetics: a conceptual approach*, 2nd ed. New York: W.H. Freeman and Co.

Russell, P.J. 2006 *iGenetics: a molecular approach.* San Francisco: Benjamin Cummings.

Snustad, D.P., and M.J. Simmons. 2006. *Principles of genetics*, 4th ed. New York: John Wiley & Sons.

Watson, J.D. 1968. *The double helix.* New York: Atheneum.

Watson, J.D., and F.H.C. Crick. 1953a. Molecular structure of nucleic acids. A structure for deoxyribose nucleic acid. *Nature* 171:737–738.

———. 1953b. Genetical implications of the structure of deoxyribonucleic acid. *Nature* 171:964–969.

INVESTIGATION 15

Restriction Endonuclease Digestion and Gel Electrophoresis of DNA

Two important procedures in the development of modern molecular genetics are restriction endonuclease digestion and agarose gel electrophoresis of DNA. These two procedures allow for the cutting of any DNA into fragments of specific size and their separation by molecular weight in an electrical field. These procedures, along with the use of polymerase chain reaction (PCR), reverse transcription, transformation with plasmids, and cloning (to be discussed in Investigations 16 and 17), form the basis of molecular genetics.

Restriction endonucleases are enzymes that cut DNA at specific nucleotide sequences in either a staggered or a straight cut. For example, the restriction endonuclease Eco RI (named for the organism from which it was isolated) makes a staggered cut (arrows) in the hexanucleotide sequence

$$\downarrow$$
$$5' - GAATTC - 3'$$
$$3' - CTTAAG - 5'$$
$$\uparrow$$

to produce the fragments

$$5' - G \qquad AATTC - 3'$$
$$3' - CTTAA \qquad G - 5'$$

On the other hand, Hind III cuts the nucleotides sequence

$$\downarrow$$
$$5' - AAGCTT - 3'$$
$$3' - TTCGAA - 5'$$
$$\uparrow$$

to produce the fragments

$$5' - A \qquad AGCTT - 3'$$
$$3' - TTCGA \qquad A - 5'$$

Note that the nucleotide sequences for both Eco RI and Hind III are palindromic sequences (nucleotide sequences that are read the same in both directions, as are the words *noon* or *eye*). Most, but not all, restriction endonucleases act on palindromic sequences. Two other characteristics of restriction endonucleases are the lengths of the recognition sequence (most are either hexanucleotides or tetranucleotides) and the effect of methylation of the recognition sequence on the enzyme activity of some restriction endonucleases.

163

If a restriction endonuclease recognizes a hexanucleotide sequence, then it would be expected to cut genomic DNA at random approximately every 4100 base pairs (four different nucleotides taken to the sixth power, or 4^6). Of course, not all DNA fragments from single restriction endonuclease digestion would be exactly 4100 base pairs in length; some would be much smaller and others larger. However, the average for a particular genomic DNA should be approximately 4100 base pairs, given certain assumptions about the composition of the DNA. An endonuclease that recognizes a tetranucleotide sequence will produce an average fragment of $4^4 = 256$ base pairs. This information can be used to determine the approximate size of the DNA fragment(s) with which you want to work.

DNA methylation (a process in which a methyl group is attached to either an adenine or a cytosine base) influences whether or not a specific restriction endonuclease will recognize its recognition sequence in order to cut the DNA. The DNA of some plasmids is protected by methylation from digestion by restriction enzymes. If the DNA with which you are working is methylated, then restriction endonucleases that are not affected by methylation should be used.

Agarose gel electrophoresis is based on the principle of separating molecules by their attraction to an electrical charge. Because DNA contains a negative charge resulting from the phosphate groups linking the deoxyribose backbone, it will migrate toward the positive pole when placed in an electrical field. Agarose is a highly purified form of agar and when solidified will make a network through which DNA must move. The result is that large DNA fragments will migrate more slowly through the agarose, whereas smaller DNA fragments will migrate more rapidly. These different rates allow for the separation of a mixture of DNA fragments by their molecular weights. The molecular weight of a particular DNA can be used to determine the approximate number of nucleotides in that DNA. (*Note:* Other factors such as linear versus circular DNA, supercoiling, and ionic strength of the buffer can also influence the migration of DNA. For a further discussion, see Maniatis et al., 1982; Sambrook et al., 1989.)

As we will see in Sections VI and VII of this investigation, restriction endonuclease fragments can be used for mapping DNA (genes), and **restriction fragment length polymorphisms (RFLPs)** can serve as genetic markers of human genetic diseases.

OBJECTIVES OF THE INVESTIGATION

Upon completion of this investigation, you should be able to

1. **outline** a procedure for using restriction endonucleases to digest DNA,

2. **describe** a protocol for using agarose gel electrophoresis to separate the fragments of DNA produced by restriction endonuclease digestion,

3. **estimate** the approximate size of a DNA fragment by comparison to a known DNA molecular size (weight) standard,

4. **describe** how restriction endonuclease-produced fragments can be used for mapping the DNA of an organism, and

5. **solve** problems involving the use of restriction fragment length polymorphisms (RFLPs) as markers of a genetic disease in a human kindred.

Materials needed for electrophoresis: (Instructors should refer to the *Instructor's Manual* for detailed instructions, recipes, cost-cutting ideas, and vendors.)

For the class:

    1 analytical balance

    1 bottle of agarose[1]

---

[1] Sigma Chemical Company, P.O. Box 14508, St. Louis, MO 63178-9916, or Life Technologies, Inc., GIBCO BRL, 3175 Stanley Rd. Grand Island, NY 14072.

1 roll of autoclave tape

1 ultraviolet (UV) light transilluminator or handheld UV mineral light

1 vortex mixer

1 waste container for ethidium bromide solution

1 bottle of ethidium bromide stock solution (10 mg/ml). *Note:* If staining with methylene blue instead of ethidium bromide, one bottle of 1.0% aqueous methylene blue stain

1 roll of Saran Wrap

incubator, water bath, or Dri Block with holder for Eppendorf tubes adjusted to 37°C; a second water bath or Dri Block adjusted to 50°C

different restriction endonucleases (examples: Eco RI and Hind III)[1]

different 10× stock endonuclease digestion buffers

1 tube of gel dyes

supply of distilled water

DNA molecular size markers[2]

If the class is not restriction-cutting its own Lambda DNA:

supply of precut Lambda DNA[3] (for example, EcoRI, Hind III, and EcoRI-Hind III double digest) adjusted to a concentration of 1.5 µg DNA per student sample (15 µl). Each individual student sample should then be combined with 5 µl gel dyes to make a 20 µl sample.

For each group of students (as many as five to seven per group depending upon the capacity of the gel apparatus):

1 agarose gel apparatus with combs and glass plate[4] (We recommend a 14-cm gel apparatus for the best visualization of the DNA fragments.)

1 power supply

1 250-ml flask

1 weighing boat

1 spatula

2 liters of chilled electrophoresis buffer

1 100-ml graduated cylinder

1 magnetic stirring hotplate and stirring bar

1 insulated "hot" glove or towel for handling hot material

1 Pipetteman, Finnpipette or similar pipette capable of handling as little as 1 µl of solution

1 Pipetteman, Finnpipette or similar pipette capable of handling as little as 20 µl of solution

supply of pipette tips

supply of 0.5-ml Eppendorf tubes

1 unmethylated Lambda DNA solution (concentration 1.5 µg/12 µl)

1 Eppendorf tube of gel dyes

1 plastic or glass container approximately 8 × 8 × 2 inches

1 pair of plastic or latex gloves

---

[2] We recommend Sigma Chemical Company's Wide-Range DNA Marker (16 fragments ranging from 50 bp to 10,000 bp) cat. no. D 7058.

[3] Eco RI, Hind III, double digest, and other restriction-cut Lambda DNA can be purchased from the Sigma Chemical Company.

[4] Life Technologies, Inc., GIBCO BRL, 3175 Stanley Rd. Grand Island, NY 14072; Edvotek, P.O. Box 1232, West Bethesda, MD 20827.

I.   RESTRICTION ENDONUCLEASE DIGESTION OF LAMBDA DNA

Lambda is a bacteriophage that has a genome approximately 50 kilobases (50 kb) in length and that consists of linear, double-stranded DNA. Because of its size and the fact that it can be obtained commercially in a nonmethylated form, Lambda DNA is ideal for demonstrating restriction analysis of DNA. Today you will analyze restriction fragments of Lambda DNA cut by at least two different restriction endonucleases—for example, Eco RI and Hind III—and a double digest (both Eco RI and Hind III). Your instructor may choose different (or include additional) endonucleases. If so, he or she will inform you of which combination(s) of restriction endonucleases you will use in your group. The quantities referred to in the following text are for a large-size gel (14 cm). If you are using a smaller gel apparatus, quantities may be adjusted accordingly.

Prior to class your instructor prepared the following solutions.

10× digestion buffer. (Different restriction enzymes may require different digestion buffers; be sure to use the correct one. Directions for making the digestion buffers come with the restriction enzymes.)

An aqueous Lambda DNA solution with a concentration of 1.5 μg DNA/12 μl.

1. Your instructor will assign each person to use one of the available restriction enzymes or a combination of two restriction enzymes (double digest). Different members of your group should be assigned different enzymes so that everyone will see the different fragment patterns after gel electrophoresis. Make sure to record the precise restriction enzyme or combination of enzymes that is assigned to you.

2. Use a Pipetteman to transfer 12 μl of the Lambda DNA solution to a 0.5-ml Eppendorf tube.

3. Next, place into the Eppendorf tube 1.5 μl of the 10× digestion buffer for your restriction enzyme and mix on a vortex mixer or by flicking with your finger.

4. Restriction endonucleases are stable when stored in a 30–50% glycerol solution at −20°C but are unstable in a working digestion buffer. (They are also very expensive.) For these reasons, your instructor will prepare a working enzyme solution while you are doing steps 2 and 3. In a working enzyme solution, 1 μl contains 1 unit of enzyme activity. (One unit is defined as the amount of enzyme necessary to digest 1 μg of Lambda DNA in 1 hour at 37°C.)

5. Your instructor will transfer 1.5 μl of one of the assigned restriction enzymes to the Eppendorf tube of each member of your group. (Use a new pipette tip for each transfer.) For the double digest, use 1 μl of each of the two restriction endonucleases. After addition of the enzyme, mix the contents and incubate at 37°C for 1 hour. *While the DNA is being digested, start preparing your agarose gel following the procedure outlined in Section II.*

6. At the end of 1 hour, remove your Eppendorf tube from the 37°C incubator or water bath and add 5 μl of gel dyes, then mix. If some of the mixture remains on the side of the tube, gently tap the tube on the table to force the solution down to the bottom of the tube.

7. Proceed with the gel electrophoresis procedure (Section III).

II.   PREPARING THE GEL

1. Weigh 2.0 g agarose in the weighing boat and carefully add the agarose to the 250-ml flask, making sure that none of the agarose touches the sides of the flask.

2. Measure 100 ml of electrophoresis buffer in a 100-ml graduated cylinder and then carefully pour the buffer into the 250-ml flask containing the agarose. This will make a 2.0% agarose gel. Next, add the magnetic stirring bar to the flask and place the flask on the magnetic hotplate. Turn the hotplate to high and the stirrer to a position such that the solution is being completely mixed. Heat the mixture until it starts to boil. The mixture will appear a cloudly white at first, but will be clear when it is ready to use. (Your instructor might prepare the agarose for the entire class and then distribute the appropriate amount of melted agarose to each group.)

3. While the agarose is heating, rinse the electrophoresis apparatus with distilled water if it appears dirty, and then dry it with paper towels. Be careful not to damage the thin platinium wire electrodes that are at the bottom of each end of the apparatus.

4. Observing the inside of the electrophoresis apparatus, you will see a central platform, which is open on each end. This area will hold the gel when it is poured. Using the tape that is provided, tape the two open ends of the gel platform. Rub your fingernail across the tape where it meets the platform to ensure a tight seal. (Some electrophoresis chambers have a platform or gel former that is removable so that the taping procedure can be performed outside the chamber. Other electrophoresis chambers have the platform at the bottom of the apparatus and have the ends sealed with plastic or metal inserts. If this is the case, your instructor will demonstrate how to set this up.) We recommend that autoclave tape be used because the glue used on this type of tape can withstand the high temperature of the melted agarose and hence not leak during the gel-forming procedure. *Note:* Your instructor may have you place a glass plate on the bottom of the gel former for easier handling of the gel during later steps.

5. After taping the ends of the platform, place the Teflon spacer comb on the platform using the slots provided. Place the comb such that the spacers (teeth) are as far away from one edge of the platform as possible. In addition, make sure the comb teeth do not touch the bottom of the gel platform. After the agarose is poured and the gel is solidified, the comb will be removed. The spacers will create wells to which DNA samples can be added. Most combs have eight spacers, so that eight wells will be present in the gel; other combs may have as few as six teeth or as many as ten teeth.

6. Make sure that the electrophoresis apparatus is resting level. When the agarose just starts to boil, remove it from the hotplate using an insulated glove or towel, and carefully pour it into the center of the gel platform. Some gel formers require that the agarose be cooled to avoid warping the plastic. If this is the case, your instructor will inform you. Pour enough melted solution so that the agarose is approximately 1 cm thick. Make sure there are no strong air drafts that would cause ripples in the gel, and then allow the gel to solidify. (The lid of the electrophoresis chamber can be gently placed over the gel to prevent air currents from disturbing the gel as it solidifies.) The gel will become opaque when it has solidified. This will require approximately 15–20 minutes.

7. Once the gel is solidified, remove the tape from the edges of the platform and fill the chamber with electrophoresis buffer so that it completely covers the gel by approximately 1 cm. Allow the gel to sit in the buffer for a minute or two, and then remove the comb by pulling up evenly on both ends of the comb. The buffer should fill the wells when the comb is removed. If a bubble is present in a well, use the Pipetteman to blow buffer into the well forcing the bubble out. The gels may sit several hours in this state. *Note:* If the comb is removed before the gel is sufficiently solidified, the wells may be damaged, resulting in sample leakage from the wells.

III. LOADING THE GEL AND GEL ELECTROPHORESIS

1. Your DNA sample has been combined with approximately 5 µl of the gel dye solution, for a total volume of approximately 20 µl, and is in your 0.5-ml Eppendorf centrifuge tube. (*Note:* Because of time constraints, your instructor may provide you with restriction-cut Lambda DNA instead of having you do the restriction digest.[5]) In this case, your instructor will provide you with Lambda DNA cut with EcoRI, Hind III, and an EcoRI-Hind III double digest DNA, as well as possibly other restriction-cut DNAs. Heat all samples in a Dri Block (water bath) for 2 minutes at 50°C to prevent interactions of Cos sites in the Lambda DNA. Then remove the samples to room temperature, and while your samples cool, adjust your Pipetteman to 20µl and attach a clean tip. (Your instructor will demonstrate the proper use of the Pipetteman.) Depress the plunger on the Pipetteman to the first stop; then, place the tip to the bottom of the DNA sample and slowly release the plunger.

---

[5]   Both Eco RI and Hind III restriction-cut Lambda DNA can be purchased from Sigma Chemical Company.

This action will allow 20 µl of the DNA sample to be drawn into the pipette tip. Starting from the second well of the gel and proceeding toward the other end for each member of your group, position the tip of the pipette at the top of a well and slowly depress the applicator button to the first stop position. Because the gel dyes contain compounds denser than the buffer, the DNA sample will flow to the bottom of the well. *Do not depress the applicator button to the blow-out position;* doing so would force air into the well, causing the sample to be ejected. *Your instructor will place a known DNA standard consisting of various-size DNA fragments into the first well (lane) of the gel and possibly the last well if sufficient space is available.*

2. When all members of your group have placed their samples into the wells, put the chamber lid onto the electrophoresis chamber. Attach the black wire to the negative pole and the red wire to the positive pole of the chamber, and then attach the other ends to the respective poles of the power supply. *Note:* DNA is negatively charged; therefore, the positive pole should be at the far end from the sample wells.

3. Set the power supply at 200 V and turn it on for 5 min. to "stack" the DNA against the wall of the gel well, then reduce the voltage to 75 V for the remaining gel run. *Warning: You are dealing with an electrical source. Do not touch the buffer or unit while the power is on.* Stacking the DNA with a high voltage produces sharper bands during electrophoresis. As the sample migrates through the gel, you will begin to see a blue dye band and a blue-green dye band. Let the gel electrophoresis proceed until the first band is at least two-thirds of the way through the gel and the second one is one-third of the way through. This should require between 2 and 4 hours depending on the gel size. (A 14-cm gel will require approximately 4 hours running time at 75 V. Generally, for 14-cm gels the fastest migrating dye band can be allowed to migrate to within 1 to 2 cm of the bottom of the gel.) Increasing the voltage will decrease the amount of time that is required for migration through the gel, but it will also decrease the resolution of the DNA bands and increase the temperature of the running buffer. Approximately halfway through the running time, stop the electrophoresis and replace the gel-running buffer with fresh chilled buffer, then restart the electrophoresis. (This is especially important if the buffer becomes hot. If the buffer remains cool or just warm, no buffer change is required.)

IV. STAINING THE GEL

1. Place approximately 400 ml of distilled water into a container and add 100 µl of ethidium bromide stock solution. Gently swirl the container to mix the stain. *Always wear gloves when working with ethidium bromide because it is a mutagen and potential carcinogen. Your instructor may prefer to stain your gels for you. Note:* If staining with methylene blue, see step 9.

2. Because it is desirable to flip the gel over for viewing and photographing, it becomes easy to confuse which is the first and the last lanes of DNA. To avoid this potential problem and to mark each group's gel, asymmetry can be introduced into the gel by making a small notch at one side of the gel bottom with a spatula.

3. Carefully lift the gel from the chamber platform and transfer to the ethidium bromide solution. *Note:* Agarose gels are flimsy and can easily tear, so handle them with care.

4. Place the container on a shaker at low speed for 30 minutes, or if a shaker is unavailable, manually rock the tray occasionally.

5. After the 30-minute staining period, pour the ethidium bromide solution into the waste container provided, being careful not to let the gel slide out of the container.

6. Gently pour approximately 400 ml of distilled water over your gel and wash the gel for 20 minutes at low speed on the shaker. After the 20-minute wash, discard the distilled water into the waste container. Repeat this process two additional times for a total of 60 minutes wash time.

7. Using a spatula, carefully lift one end of the gel and slide your fingers under it. Lift the gel out of the container and flip it over; then, place it on a piece of Saran Wrap. Since the DNA was at the

bottom of the well, flipping the gel over places the DNA closest to the camera for observing and photographing. (Be sure to wear disposable gloves for this procedure.)

8. While it is still on the Saran Wrap, place the gel on the ultraviolet transilluminator, which has been positioned behind a plexiglass shield to reduce your exposure to ultraviolet light. *Put on plastic safety glasses to protect your eyes further from the UV light, and then turn on the transilluminator.* Ethidium bromide will fluoresce orange anywhere it has intercalated with DNA. If a UV transilluminator is not available, a handheld UV mineral light can serve as a somewhat satisfactory substitute. For viewing a gel with a handheld illuminator, place the Saran-Wrapped gel on the illuminator's glass plate and view directly. You may wish to remove the Saran Wrap from the top of the gel for better viewing. Your instructor will demonstrate how to use the camera and photographic apparatus, if you are photographing the gel. If not photographing the gel, then quickly draw the banding pattern(s) in the spaces provided in Figure 15.1.

9. For **methylene blue** staining, place approximately 400 ml of distilled water into a container and add 10 ml of methylene blue stock solution. Gently swirl the container to mix the stain. Place the gel into the staining solution using the procedure given for ethidium bromide and stain for 20–30 minutes. After the staining period, remove the methylene blue solution and replace with several changes of clean water. Continue to destain the gel with water until the DNA bands become evident, and then wrap the gel in Saran Wrap and view on a standard (visible) light box.

Which restriction endonuclease did you use, and how many DNA bands did you observe

in your restriction-cut DNA? _____

How did your DNA banding pattern differ from those of the other restriction endonucleases?

_____

_____

What explanation can you provide for the differences in the migration of the bands? _____

_____

_____

Did any of the bands appear the same in the two restriction-cut DNAs? _____

If so, what does this similiarity suggest to you? _____

_____

V. DETERMINATION OF DNA RESTRICTION FRAGMENT SIZE

Figure 15.2 is a photograph of a typical student gel produced by the procedure outlined above. In this photograph, lanes 1 and 6 have marker DNAs with sizes ranging from 10,000 base pairs at the top of the gel down to a couple of hundred base pairs at the bottom of the gel. Lane 2 has Hind III digested Lambda DNA; lane 3, double digested DNA; lane 4, Eco RI digested DNA; and lane 5, Lambda DNA digested with an unknown restriction endonuclease. Can you identify the restriction endonuclease which was used to cut the lambda DNA in lane 5? If so, place your answer below and explain why you came to this conclusion.

_____

_____

_____

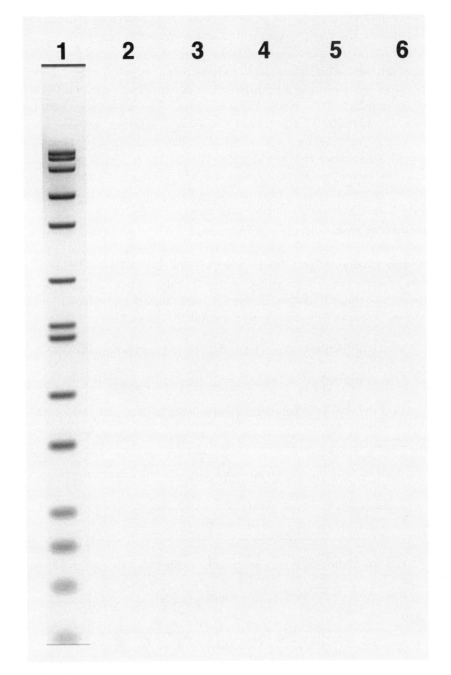

FIGURE 15.1. A diagram with DNA size standards (markers) and spaces for recording the banding patterns in student gels. This is the same DNA size standard that is used in the gel in Figure 15.2. The size (in base pairs) of the DNA size standards is as follows from top to bottom: 10,000 bps, 8000, 6000, 4000, 3000, 2000, 1500, 1400, 1000, 750, 500, 400, 300, 200, (100, 50 not observable). (The dark line immediately below the lane number marks the upper edge of the well and can serve as a reference point for measuring the distance to individual bands.)

The size in base pairs (hence the approximate molecular weight) of the DNA fragments separated by the agarose gel electrophoresis can be determined by comparison of their mobility during electrophoresis with the mobility of the known DNA markers loaded into the first lane (and possibly the last lane) of your gel. The procedure for this determination is relatively easy and involves measuring

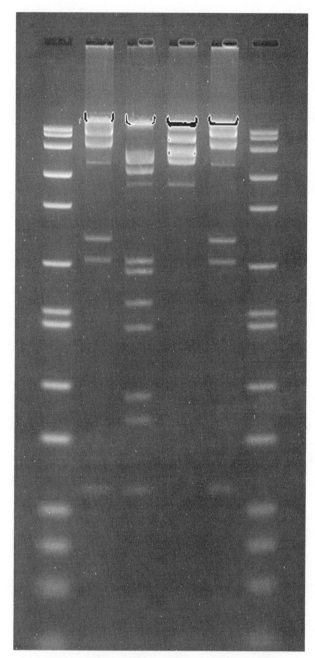

LANE

FIGURE 15.2. Photograph of a typical 14-cm electrophoresis gel of several restriction-cut Lambda DNAs. Lanes 1 and 6 contain a DNA size standard with marker DNAs ranging from 10,000 base pairs (bps) to 200 bps. The size (in base pairs) of the DNA size standards is as follows from top to bottom: 10,000 bps, 8000, 6000, 4000, 3000, 2000, 1500, 1400, 1000, 750, 500, 400, 300, 200, (100, 50 not observable). Lane 2 contains Hind III cut DNA; lane 3 Hind III-Eco RI (double) cut DNA; lane 4 Eco RI cut DNA; and lane 5 Lambda DNA cut by an "unknown" restriction enzyme.

the distance (in mm) from the leading wall of the well to the center of a given band for each of the known DNA markers, and then plotting the log of base pair size versus the distance traveled during electrophoresis. This is very easily done by graphing your data on semilog graph paper. (Semilog graph paper has a log scale on the y-axis and a normal scale on the x-axis.) A sheet of semilog paper is located at the back of the book, page 306.

1. Determine the distance traveled for each band of the DNA marker lane (lane 1) and record

    your data here. _____

    _____

    _____

2. Using a photocopy of the semilog paper on page 306, plot the length of each DNA marker fragment in base pairs on the log scale (y-axis) and the distance traveled in mm for each marker on the x-axis. Note that this plot is a more-or-less straight line between 100 bp to 4000 bp fragments but not with the larger fragments. Do not try to draw a straight line between these larger fragments.

3. Determine the distance traveled for each band generated in the Hind III digest and record

    your data here. _____

    _____

    _____

4. Determine the distance traveled for each band generated in the double digest and record

    your data here. _____

    _____

    _____

5. Determine the distance traveled for each band generated in the Eco RI digest and record

    your data here. _____

    _____

    _____

6. Utilize the graph constructed in steps 1 and 2 above to determine the size in base pairs for each fragment generated with each restriction enzyme. This is very simply done by finding the distance traveled on the x-axis, then moving up vertically to the line, then horizontally to the left (y-axis) to see what size fragment would generate that point.

7. Complete the following chart using the data collected in step 6.

Enzyme	Hind III	Double Digest	Eco RI
No. of Bands			
Size in Base Pairs (from largest to smallest)			

VI. RESTRICTION MAPPING

The different DNA fragments generated by different restriction endonucleases can actually be used to map DNA. For example, a specific DNA sequence 2000 nucleotides in length can be cut by either Eco RI or Hind III and generate the DNA fragments as illustrated in the following gel diagrams:

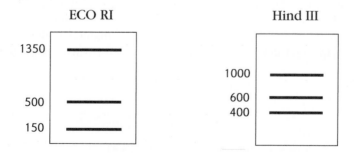

In this case, Eco RI produced three fragments (1350, 500, and 150 nucleotides in length) and Hind III three fragments (1000, 600, and 400 nucleotides in length). These fragments alone do not provide enough information to map the DNA sequence. Instead, what is required is a double digest, where the

Eco RI fragments are redigested with Hind III and the Hind III fragments redigested with Eco RI to generate the following gel patterns:

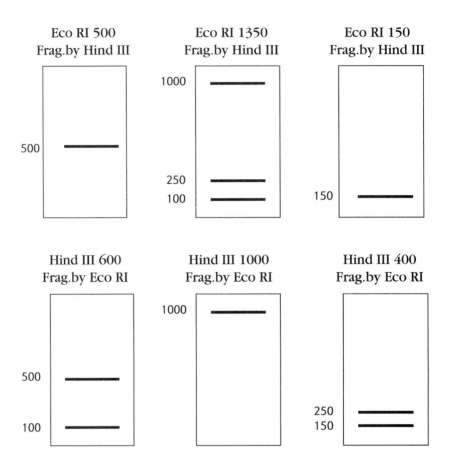

Because the Eco RI 500 and the Eco RI 150 fragments remain the same, there are no Hind III restriction sites in either of these fragments. On the other hand, the Eco RI 1350 fragment is cut into 1000, 250, and 100 fragments by Hind III, which indicates that there are two Hind III sites in this fragment. The Hind III 600 fragment is cut once by Eco RI to produce 500 and 100 fragments. This indicates that the Eco RI 500 fragment overlaps the Hind III 600 fragment. Conversely, the Eco RI 1350 fragment overlaps by 100 base pairs the Hind III 600 fragment and overlaps both the 1000 and 400 Hind III fragments. Because the Eco RI 1350 fragment has a 250-nucleotide fragment, it also must overlap with the Hind III 400 fragment. The Hind III 400 fragment contains a 150-base-pair fragment and so must overlap with the Eco RI 150-base-pair fragment. Thus, a restriction map can be contructed as follows:

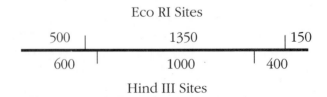

The diagram above represents a simple example of restriction mapping using isolated fragments for a redigestion. However, in practice, this approach is usually not followed. Instead, three reaction tubes are set up—one containing Eco RI, one with Hind III, and the third containing both Eco RI and

Hind III, so that a simultaneous double digest occurs. The following lists show what the fragment size might look like for a DNA approximately the size of Lambda:

Eco RI	Eco RI and Hind III	Hind III
21200	21200	23100
7450	4400	5350
6200	4300	5100
5800	3700	4750
350	2700	2700
	1900	
	1400	
	1050	
	350	

The Eco RI 21,200 fragment must overlap with the Hind III 23,100 fragment, so that in the double digest a 1,900 fragment is produced ($23,100 - 21,200 = 1,900$). This result means the next Eco RI fragment also contains that 1,900-nucleotide sequence, so that the 1,900 fragment plus another double-digest fragment must equal one of the Eco RI fragments. This turns out to be the 4,300 fragment; 1,900 plus 4,300 equals the 6,200 Eco RI fragment. The next Hind III fragment also contains the 4,300 base pairs, so 4,300 plus a double-digest fragment must total to equal a Hind III fragment.

Use scrap paper and map this DNA; then, draw your map in the space provided below and indicate where each of the restriction sites is.

VII. MAPPING THE HUMAN GENOME AND RESTRICTION FRAGMENT LENGTH POLYMORPHISMS (RFLPS)

In the 1980s, an ambitious project to sequence the human genome completely was initiated by several researchers (White and Lalouel, 1988). During the 1980s, other researchers started participating either directly or indirectly in this project, a fact that unfortunately resulted in duplication of efforts and waste of valuable research funds. To avoid this problem, in 1988, a concerted effort to coordinate the work of individual researchers was established by the U.S. government. Dr. James Watson was named the first director of this effort and was to act as a liaison with the government. Currently, Dr. Francis Collins is the project director. The total cost of the project was approximately $3 billion. The value of the project, however, is immeasurable in terms of the scientific and medical knowledge that has been, and will continue to be, obtained. (The human genome project has now been completed and is proving to be one of the most valuable and productive projects ever undertaken by humans.)

The initial part of the project created a complete human genomic library of restriction-endonuclease-produced DNA fragments. Each of these DNA fragments has a unique nucleotide sequence and was mapped by restriction endonuclease digestion. Once this mapping was done, small portions of each fragment could have the nucleotide sequence determined directly by a process known as **DNA sequencing**. Because the initial DNA library was made by using only a partial restriction endonuclease digestion, different DNA fragments overlapped somewhat. This overlapping allowed for moving from one partially overlapped fragment to another during the restriction mapping and sequencing process, a procedure known as **chromosome walking**.

An important result of this research has already emerged in the detection of **restriction fragment length polymorphisms (RFLPs)**, regions of DNA (which are probably not true genes) that show large variation in endonuclease restriction sites from one individual to another. The significance of this development is that it is now essentially possible to distinguish any one individual from another by comparing a large number of different RFLP patterns. This is a process termed **DNA fingerprinting**.

DNA fingerprinting, because of its accuracy, has in recent years been introduced into forensic medicine. For example, a corpse can be identified by comparing its DNA to that of possible relatives. DNA fingerprinting has been used extensively in certain South American countries where former dictators had separated parents and their infant children. By DNA fingerprinting, parents and grandparents could often be reunited with their children and grandchildren. DNA fingerprinting has been used in identification of rape suspects, with as little as 5 μg of semen providing a sufficient quantity of DNA for analysis, even from corpses as old as 3 months. Similarly, DNA isolated from bone fragments has been used to analyze and identify the remains of the last Russian czar and his family (see Ivanov et al., 1996). The FBI has initiated pilot projects in which all convicted felons are DNA fingerprinted as a method of future identification. These examples represent the way in which this technology has been applied in a positive fashion. However, some politicians have suggested that all newborn children should be subjected to DNA fingerprinting and the results used on a national scale as a potential method of identification. This and similar suggestions by others raise serious ethical questions as to how this DNA technology should be applied.

Another way in which RFLPs have been used in medicine is as genetic markers for serious genetic diseases that cannot be detected by other means. Human genetic researchers are now using RFLPs (which are linked to a gene causing a genetic disease) as a marker for that particular disease; by following the way those RFLPs are inherited in a family pedigree, they can predict heterozygous carriers of the trait. Two of the first genetic diseases to be traced in this way were Huntington's disease (an autosomal dominant) and cystic fibrosis (an autosomal recessive).

Figure 15.3 is an illustration of a human pedigree showing three generations of a family in which one person (individual 7) has the autosomal recessive condition, cystic fibrosis (CF). The parents of individual 7 are concerned about the prospects that their unborn child (individual 8) will also have CF. They seek evidence from a medical genetics center about the transmission of CF in their kindred. Figure 15.3 also illustrates the results of an RFLP analysis completed on the DNA of members of the kindred. Using the information from Figure 15.3, identify the RFLP genotypes of individuals 1 to 7 in the pedigree (for example, individual 1 has the genotype AC).

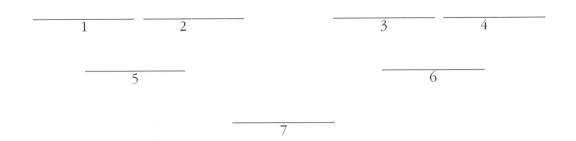

Now, using this information, and keeping in mind that individual 7 has CF, trace the pattern of inheritance of the CF allele through the three generations of this kindred. The parents of individual 7, for example, had to be carriers of the CF allele, but from which of their parents did they inherit the CF gene? Assume that there is no recombination between the RFLP marker and the CF gene. Now, let us attempt to predict the RFLP and CF genotypes possible in individual 8, the fetus. Cells from individual 8 could be obtained by amniocentesis or chorionic villus sampling (CVS). DNA from such cells can be analyzed using restriction enzymes and electrophoresis to determine the RFLP constitution of the

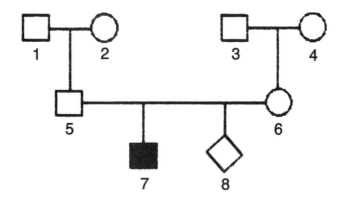

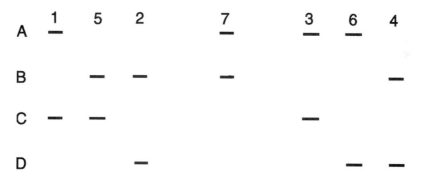

FIGURE 15.3. Human pedigree and associated RFLP markers as determined by electrophoresis. Use the information from this illustration to determine the RFLP genotype of individuals 1–7 and to trace the transmission of the cystic fibrosis gene from grandparents (1, 2, 3, 4) to parents (5 and 6) to affected child (7).

fetus. Assuming that *no* recombination occurs between the CF locus and the RFLPs, make a diagnosis relative to CF for each of the following fetuses.

Fetus RFLP constitution	Genotype relative to CF
AC	_____
BD	_____
CD	_____
AB	_____

You ought to have been able to predict which of these RFLP constitutions were associated with homozygous normal, heterozygous, and homozygous recessive genotypes vis-à-vis cystic fibrosis.

The predictions you have just made were based on the assumption that no recombination occurred between the RFLP markers and the gene locus for CF. If, on the other hand, recombination occurs between the CF locus and the RFLP markers, how would this affect the accuracy of your predictions?_____

_____

_____

_____

Compare the accuracy resulting when the CF locus is closely linked to the marker as opposed to being much less tightly linked to the marker. _____

_____

_____

_____

_____

APPENDIX

Ethidium bromide stock solution (10 mg/ml)

10× electrophoresis buffer—stock solution (8 liters)

387.2 g Tris 7–9	Conc. = 200 mM
217.7 g NaOAc · 3H$_2$O	Conc. = 200 mM
59.5 g Na$_2$EDTA · 2H$_2$O	Conc. = 20 mM

Bring to 7 liters with ddH$_2$O

pH to 7.8 with 120–130 ml glacial acetic acid
Bring to 8 liters with ddH$_2$O

Gel dyes

5.0 g Ficoll 400 (Sigma Chemical Co.)
5.0 g glycerol (Eastman)
10.0 ml 0.1% bromophenol blue, xylene cyanol
FF (another dye)
fill to 50 ml with ddH$_2$O

REFERENCES

BILLINGS, P.R., ed. 1992. *DNA on trial: genetic identification and criminal justice.* Plainview, NY: Cold Spring Harbor Laboratory Press.

BSCS and AMA. 1992. *Mapping and sequencing the human genome: science, ethics, and public policy.* Colorado Springs, CO: The Biological Sciences Curriculum Study.

BUDOWLE, B., H.A. Deadman, R.S. Murch, and F.S. Baechtel. 1989. DNA fingerprinting: under investigation in the FBI lab. *The Philter* 22(1):1–8.

COOPER, N.G., ed. 1992. *The human genome project. Los Alamos science* No. 20. Los Alamos, NM: Los Alamos National Laboratory.

DRLICA, K. 2003. *Understanding DNA and gene cloning: a guide to the curious,* 4th ed. New York: John Wiley & Sons.

IVANOV, P.L., et al. 1996. Mitochondrial DNA sequence heteroplasmy in the Grand Duke of Russia Georgij Romanov establishes the authenticity of the remains of Tsar Nicholas II. *Nature Genetics* 12(4):417–420.

JONES, P. 1996. *Gel electrophoresis: nucleic acids: essential techniques.* New York: John Wiley & Sons.

LEWIN, B. 2004. *Genes VIII.* Upper Saddle River, NJ: Pearson Prentice Hall.

MANIATIS, T., E.F. FRITSCH, and J. SAMBROOK. 1982. *Molecular cloning: a laboratory manual.* Plainview, NY: Cold Spring Harbor Laboratory Press.

NEUFELD, P.J., and N. Colman. 1990. When science takes the witness stand. *Scientific American* 262(5):46–53.

PATEL, D., ed. 1994. *Gel electrophoresis.* New York: Wiley-Liss.

PINES, M., ed. 1991. *Blazing a genetic trail.* BETHESDA, MD: Howard Hughes Medical Institute.

SAMBROOK, J., and D.W. RUSSELL. 2001. *Molecular cloning: a laboratory manual* (3-volume set), 3rd ed. Plainview, NY: Cold Spring Harbor Laboratory Press.

SAMBROOK, J., E.F. FRITSCH, and T. MANIATIS. 1989. *Molecular cloning: a laboratory manual,* 2nd ed. Plainview, NY: Cold Spring Harbor Laboratory Press.

Scientific American, eds. 2002. *Understanding the genome (Science made accessible).* New York: Warner Books.

STRACHAN, T., and A.P. READ. 2003. *Human molecular genetics,* 3rd ed. New York: Garland Science/Taylor & Francis Group.

WESTERMEIER, R., et al. 2005. *Electrophoresis in practice: a guide to methods and applications of DNA and protein separations,* 4th ed. New York: John Wiley & Sons.

WHITE, R., and J.-M. LALOUEL. 1988. Chromosome mapping with DNA markers. *Scientific American* 258(2):40–48.

INVESTIGATION 16

Amplification of DNA Polymorphisms by Polymerase Chain Reaction (PCR) and DNA Fingerprinting

The idea for the polymerase chain reaction was first formulated by Nobel Laureate Dr. Kary Mullis while he was having a leisurely drive home after a day's work. His thoughts, during this drive, centered on DNA replication and the enzymes involved in the replication process, most specifically on the properties of DNA polymerases (Mullis, 1990). After thinking about the properties of DNA polymerases and replication, it occurred to him that it should be possible to amplify millions of copies of a specific DNA sequence by manipulating the DNA polymerase and the requirements it has for replication. From this idea emerged the PCR technique, which has revolutionized molecular biology and all of genetics. Through the use of PCR it is now possible to amplify and analyze any DNA that can be isolated, whether that DNA is from a 120-million-year-old extinct weevil (Cano et al., 1993), a stuffed museum specimen, a corpse, or a living embryo or organism. Thus, PCR has become an integral process not only in the molecular analysis of an organism's genome and gene expression but also in medical genetics, with the detection of heterozygotes and individuals susceptible to a variety of diseases, in population genetics with the analysis of allele frequencies and changes in allele frequency (DNA markers are like any other genetic marker and are inherited as codominants in standard Mendelian fashion), in the study of evolutionary history and divergence of genes and organisms, and in forensic science with DNA fingerprinting.

Two major properties of DNA polymerase are responsible for the development of the PCR technique. First, all DNA polymerases must polymerize a new DNA molecule in a 5′ to 3′ direction using an existing 3′ to 5′ DNA template. Second, no DNA polymerase can initiate DNA synthesis on a single-strand template by itself, but rather must have a free 3′OH terminus in order to start polymerization (i.e., some type of primer is required). In PCR, double-stranded DNA is converted to single-stranded DNA by heating to denaturation conditions (usually about 94°C). Oligonucleotide primer DNA of a specific sequence (usually 15–30 nucleotides in length) is then allowed to hybridize to complementary sequences on the single-stranded DNA at reduced temperatures (usually between 33°C and 65°C). (*Note:* For both complementary strands of a specific DNA to be replicated, a primer for each 5′ end of those strands must be used.) The 3′ end of the primer provides a 3′OH terminus on which the DNA polymerase can start synthesis of a new strand of DNA. This synthesis is usually carried out at 72°C. The DNA polymerase used in the reaction is isolated from either *Thermus aquaticus* or *Thermus thermophilus* (both eubacteria isolated from hot springs). The DNA polymerase (Taq or Tth) isolated from these organisms shows much greater thermostability than the polymerases isolated from other organisms. Herein lies the basis of the third major feature of the polymerase chain reaction: Radical changes in temperature will not destroy the polymerization activity of the enzyme by denaturation. This feature allows multiple cycles of DNA replication to occur by repeatedly switching from a lower temperature at which replication can occur, to a higher temperature at which the replicated DNA is denatured to single-stranded DNA, followed by a switching back to a lower temperature, which allows new primers to anneal and a new round of replication to occur. The result is that, with repeated cycles of this temperature switch, an amplification of the DNA specified by the primers occurs. Figure 16.1 is a schematic representation of this process.

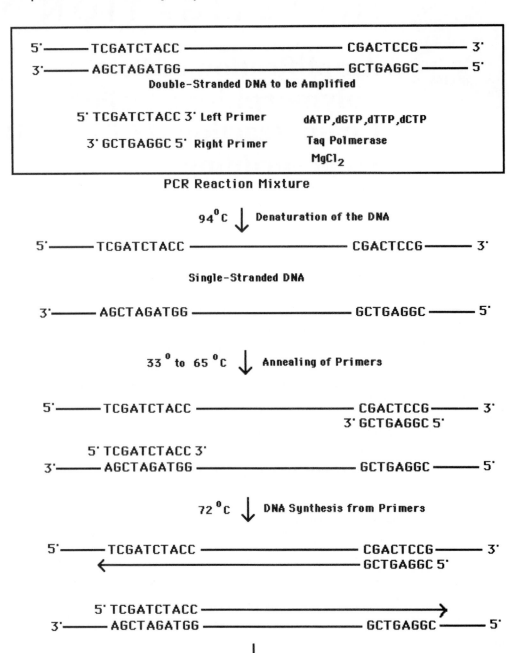

FIGURE 16.1. The events of PCR are outlined, beginning at the top of the diagram, with the DNA to be amplified and the "raw materials" needed for PCR to occur. This step is followed by denaturing the DNA at 94°C, annealing of primers at 33–65°C (depending on the primers used), and synthesizing complementary DNA at 72°C.)

In this investigation, you will use PCR to amplify and detect DNA polymorphisms in different individuals and, thus, carry out a type of DNA fingerprinting. DNA polymorphisms are different forms of the same type of DNA. These different forms can be detected in several different ways. For example, a change in a single base within a given DNA sequence can result in the production or loss of a restriction endonuclease site, causing a change in RFLP (restriction fragment length polymorphism) pattern (see Investigation 15). Alternatively, an increase or decrease in size of a DNA sequence between two flanking primers used in PCR will result in different-sized PCR products, which migrate differently during agarose gel electrophoresis. Likewise, two flanking primers may be complementary to two distinctly different DNA regions (for example, multigene families) and, thus, produce different-sized PCR products. In the latter case, the number of different DNAs that a set of primers recognizes should be, at least to a certain extent, a function of the size of the primers. The smaller the primers (number of nucleotides long), the more likely it is that they will hybridize to different DNAs. The larger the primers, the more unique they are, and the more likely they will bind to a single, unique DNA. For this investigation, your instructor will choose a set or sets of primers, such as primers for the human D1S80 "gene" locus, other minisatellite primers, or RAPD primers.

The source of the DNA for this investigation will be human DNA extracted from your own cells; thus, you will be DNA fingerprinting yourself, although, the procedures described here may be used with DNA isolated from any organism, provided that the proper complementary primers are used. Your instructor may wish to use a nonhuman DNA source and make an "opened-ended" investigation in which you investigate the PCR pattern in one or more different species.

OBJECTIVES OF THE INVESTIGATION

Upon completion of this investigation, you should be able to

1. **outline** a procedure for conducting the polymerase chain reaction,
2. **discuss** the principles upon which PCR is based,
3. **analyze** a human DNA fingerprint based on the D1S80 locus, and
4. **list** possible uses of DNA polymorphisms.

Materials needed for this investigation: (Instructors should refer to the *Instructor's Manual* for detailed instructions, recipes for solutions, cost-cutting ideas, and vendors for supplies and equipment.)

For each student:

1 numbered paper cup with 10 ml 0.9% saline solution
1 15-ml screw-cap tube (Tubes should be sterile or at least not contaminated with human DNA.)
1 1.5-ml microfuge tube containing 0.75 ml (750 μl) Chelex solution, at an alkaline pH
1 clean 1.5-ml microfuge tube
1 small (2 × 4 × 1/2 in.) Styrofoam sheet with holes for 0.5- and 1.5-ml tubes
1 0.5-ml (or 0.2-ml) PCR tube (red) containing 20 μl PCR mixture[1]
1 0.5-ml tube (or 0.2-ml) PCR tube (green or blue) containing 12 μl MgCl$_2$ solution
1 1.5-ml tube containing distilled (deionized) water

For each group of students:

1 ice container with ice
1 Sharpie marker
1 box of Kimwipes
1 0.5-ml tube of gel dyes
1 Dri-Block heated to 95°C
3 Pipettemen 1–5 μl, 5–50 μl, and 50–200 μl ranges
1 small container of pipette tips
1 small container for discarded tips

---

[1] PCR reaction materials can be obtained from several vendors including Carolina Biological, Promega, and Sigma-Aldrich. Sigma-Aldrich sell two PCR kits that include all of the components necessary for a PCR reaction (Taq polymerase, dNTP mix, 10X reaction buffers, and MgCl solution.) and that are easy to use. These are the Taq SuperPak™, which requires some mixing, and the JumpStart-ReadyMix-REDTaqDNA Polymerase Kit, which is premixed and ready to use.

For the class in general:

Miscellaneous:

clinical centrifuge(s) with enough space for all of the samples

microfuge

Heat Dri-Block. If Dri-Block is not available, use a pan of boiling water and sheet of thin Styrofoam with holes for 1.5-ml tube (for floating on boiling water).

ice container marked *Reaction Mixture*

dropper bottle of PCR-grade mineral oil

box of disposable gloves

supply of ice

supply of distilled or deionized water

For thermocycling:

automated thermocycler (e.g., Perkin-Elmer 2400) programmed for 3 min. at 94°C hot start; 94°C, 1 min.; 65°C, 1 min.; 72°C, 1 min.; finish up, 10 min. at 72°C

If doing manual thermocycling (see also *Instructor's Manual*):

3 Corningware 10-in.-diameter "deep-dish" pie dishes

1 gallon vegetable oil

3 hotplates (at least one with magnetic stirring capability)

3 thermometers

1 container of ice

1 package 10 1/4-in. Styrofoam plates

1 stapler and thin wire

1 stopwatch

For agarose gel electrophoresis:

3–6 electrophoresis setups (depending on the number of students)

power sources

DNA size standards (DNA markers like Sigma D-7058 or S-7025). The S-7025 DNA marker set is a 50-base-pair DNA stepladder and is recommended for determining the alleles of the D1S80 locus.)

scales

autoclave tape

agarose (such as Sigma A-9539)

insulated glove

Saran Wrap

3–6 150-ml flasks

chilled electrophoresis buffer (enough for 3–6 setups plus two changes of buffer for each setup)

For staining and viewing:

1 staining tray for each electrophoresis setup

ethidium bromide solution (10 mg/ml stock solution)

distilled or deionized water

waste container for ethidium bromide

UV transilluminator or handheld UV mineral light

plexiglass shield

safety glasses

I. PCR AND FINGERPRINTING OF THE HUMAN D1S80 LOCUS[2]

The D1S80 locus of humans (also called MCT 118) is located at the distal end of the short arm of chromosome 1, and it is not a typical gene that codes for a protein. Instead, this locus is composed of a series of tandem repeat sequences 16 base pairs (bp) long and is flanked on each side by unique DNA sequences. In different individual chromosomes, the number of repeats flanked by the unique DNA sequences varies greatly, typically from as few as 14 repeats to as many as 41 repeats (in a few cases, an even greater number). Consequently, this locus is a type of variable number of tandem repeat locus, or **VNTR**. VNTRs are what we refer to as **minisatellites**. Since different chromosomes have different numbers of these D1S80 16-bp repeats, the number of repeats is inherited in standard Mendelian fashion, as an allele of this locus. At least 29 different alleles of this locus have been recognized, with different alleles being inherited in a codominant fashion. Thus, any one individual could be homozygous or heterozygous at this locus. Because there are so many different alleles, most individuals would be expected to be heterozygous. In fact, since $n = 29$, there are $n(n + 1)/2$, or 435 possible different genotypes, most of which are heterozygotes.

Polymerase chain reaction can be used to detect and determine the type of allele or alleles present within an individual. Left and right primers are constructed for the unique DNA sequences that flank the two sides of the 16-bp repeats, and PCR is performed. Because the PCR will result in the amplification of the DNA between (and including) the primers, different alleles with different numbers of repeats will produce different-sized DNA fragments, which will separate in a predictable way during electrophoresis. Each different allele will have a different molecular weight based on the total number of nucleotides in the allele. For example, one of the more common alleles in the U.S. white population is allele number 18, which has a frequency of approximately 0.26; this allele consists of 18 tandem repeats. The number of nucleotides making up the PCR product for this allele thus consists of the number of nucleotides in the left primer plus the number of nucleotides in 18 repeats plus the number of nucleotides in the right primer. This number turns out to be 129 bp for the left end and primer and the first repeat + 16 bp × 17 remaining repeats + 32 bp for the right primer, for a total of 433 base pairs. (Note that the first repeat of this locus consists of 14 base pairs instead of the 16 base pairs that the remaining repeats have.) Each specific allele will have a specific number of base pairs, which can be determined by comparing the PCR product to a molecular weight (DNA size) standard or DNA ladder after completion of electrophoresis. Figure 16.2 is a photograph of an ethidium bromide-stained agarose gel of D1S80 PCR products from six individuals. In this gel, the leftmost lane is a DNA ladder (Lambda pstI cut DNA) followed by seven lanes with PCR DNAs from six individuals. The bottom, brightly staining, distinct band bserved in several of the lanes has a size that approximates the size for allele number 18. (Note that two different lanes possess DNA from the same individual amplified in two separate PCRs.)

Carefully observe the pattern of the DNA bands in Figure 16.2 and answer the following questions:

1. How many major bands do you observe in each lane (left to right) ? _____

_____

2. How would you interpret the presence of one versus two bands? Explain. _____

_____

_____

---

[2] Analysis of the human D1S80 (MCT118) locus is based on a kit and primers developed by Perkin-Elmer Corporation. Carolina Biological Supply Company is sublicensed to carry a similar kit for educational purposes (cat. #RG-21-1233) and is an excellent source for the materials and primers (cat. #RG-21-1506) used in this investigation. The protocol used in this investigation is based in part on materials supplied by both Perkin-Elmer and Carolina Biological Supply Co.

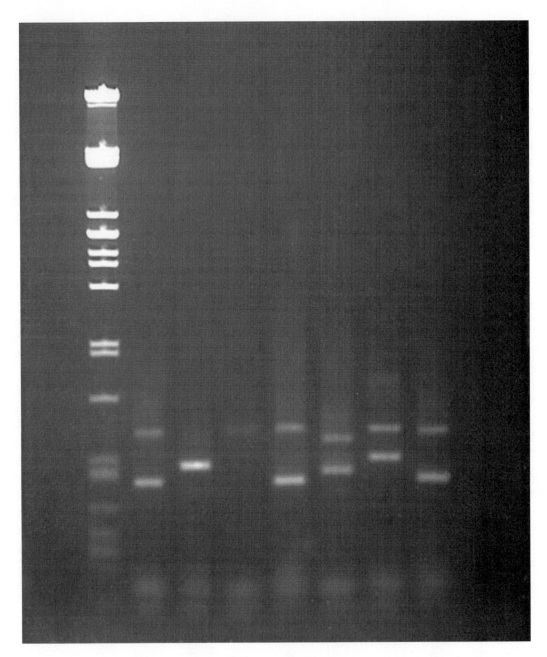

FIGURE 16.2. Photograph of a gel of PCR products showing the DNA size standards in lane 1 and PCR DNA bands for the D1S80 locus for six individuals in lanes 2–8. (One individual's DNA was amplified in two separate PCRs and the PCR products of those reactions were both electrophoresed on this gel. Can you identify which lanes came from the same person?)

3. Can you tell which two lanes possess DNA from the same individual? _____

_____

4. Can you genetically distinguish between the six different individuals represented in this gel? Explain. _____

_____

_____

(Note that in lane 4 the DNA concentration is low, making it difficult to observe all of the bands. In another gel, the DNA from this individual showed two bands: one at the same level as the very faint band seen in this gel and a second band located much higher in the gel.)

In this part of today's investigation, you will isolate DNA from yourself and prepare your own DNA fingerprint.

A.   Isolation of Cheek Cell DNA

Polymerase chain reaction will work with any DNA (not highly degraded) isolated from any source, provided that (1) there is at least 2.5 ng of DNA, (2) the DNA sample does not contain proteins and other contaminants that inhibit the Taq polymerase, and (3) nucleases that are released during isolation are inactivated before they degrade the DNA. Sources for DNA can include any tissue or blood sample, hair root follicle cells, or any body fluid that has a small number of cells in it, if those cells can be concentrated. Small numbers of nonkeratinized epithelial cells are continually shed from the lining of the mouth and will be the source of DNA for our PCR. A simple toothpick scraping of the mouth can provide enough cells for PCR; however, a more reliable method involves thoroughly rinsing the mouth with a solution, then concentrating those cells by centrifugation. Contamination of your DNA with bacteria and debris should not be a problem for the PCR, because the primers used are specific for human (mammalian) DNA, but please refrain from eating immediately before the laboratory period, because food debris can hamper the concentration of the epithelial cells.

1. At your laboratory tables, each student will be provided a small paper cup containing 10 ml of 0.9% saline solution. (*Note:* **Each cup has a number marked on it. Record and remember this number because you will be marking and identifying all of your tubes and results by this number.**) Place the solution in your mouth and very thoroughly rinse to loosen and free cells from the epithelial lining. One minute of vigorous "sloshing" should be sufficient. Return this rinse to your paper cup. Remember this is a salt solution so it will taste salty when you put it in your mouth—don't be surprised!

2. Transfer this solution from the paper cup to the 15-ml screw-cap tube located at your table, screw on the lid, and write **your number** on the tube with the marker provided. Save the cup for the time being.

3. Place this tube in the clinical centrifuge located in the laboratory. Once all students' samples have been collected, centrifuge at maximum speed for 10 minutes.

4. After the 10-minute centrifugation, remove your tube from the centrifuge, taking care not to disrupt the pellet (and cells) at the bottom of the tube. Return to your table and carefully pour the supernatant into the paper cup for later disposal down the sink. Remove all of the supernatant possible without disrupting the pellet. If the pellet starts to break up, stop removing the remaining supernatant. If only a small amount of fluid remains with the pellet, proceed to step 5; if a lot of fluid remains, recentrifuge.

5. Add the 0.75 ml (750 μl) of the Chelex resin contained in the 1.5-ml microfuge tube located at your table to the tube containing your pellet using a Pipetteman or similar type of micropipetter. (Your instructor will demonstrate the proper use of this type of pipetter.) Be sure to thoroughly resuspend the Chelex before transferring. (Chelex binds metal ions and thus inactivates certain nucleases and removes metal ions that can inhibit PCR.)

6. **Thoroughly** break up your pellet in the Chelex suspension using your pipetter, then transfer this suspension back to the 1.5-ml tube, secure the lid, and place your number on the top of the lid using a Sharpie fine-point marker.

7. Place the tube in one of the holes of a Dri-Block that has been adjusted to 95°C. Incubate for 10 minutes. This heat treatment serves to lyse the cells, denature proteins, inactivate nucleases,

and release the cells' DNA into solution. (If a Dri-Block is not available, place your tube in one of the holes of the Styrofoam sheet provided, and, once all students have positioned their samples, float the Styrofoam sheet on a pan of boiling water.)

8. At the end of the 10-minute heat treatment, transfer the tube to ice for a minute or two, then place in a microfuge. When all students have positioned their tubes, your instructor will check to ensure that all tubes are in a balanced position and then centrifuge the samples for 45 seconds. (**If a microfuge is unavailable**, your instructor will have you complete steps 6 and 7 in the 15-ml screw-cap tube. At the end of the 10-minute heat treatment, balance the tubes, place in the clinical centrifuge, and spin for 10 minutes at maximum speed.)

9. Transfer approximately 200 μl of your DNA solution to a clean 1.5-ml tube using the Pipetteman, with a clean tip. Be very careful not to transfer any Chelex. Secure the lid and write your number on the lid.

10. Place the tube on ice and proceed to section B, Preparation for PCR. *Note:* This DNA solution may be stored for days at 4°C or frozen for longer storage.

B. Preparation for PCR

1. From the ice container labeled *Reaction Mixture* obtain a 0.5-ml PCR tube containing 20 μl PCR mixture, a 0.5-ml tube containing 12 μl $MgCl_2$, and a 1.5-ml tube of water. The PCR tubes should be color coded, with the reaction mixture being in a reddish color tube and the $MgCl_2$ solution in a light green or blue tube. If colored tubes are not available, then the reaction mixture will be marked with an R on the side of the tube and the $MgCl_2$ marked with an M on the side of the tube. Do not confuse these tubes. Return to your table and place the tubes on ice. (*Note:* If your instructor plans on using an automated thermocycler for the PCR, the machine may require the use of 0.2-ml PCR tubes, in which case the reaction mixture will be in a 0.2-ml tube. Proceed with the same steps as for the 0.5-ml tubes.)

2. Write your number on the Reaction Mixture tube using the Sharpie marker. Place your number both on the lid (if possible) and on the side of the tube. You may need to dry the tube carefully with a Kimwipe. Place the tube into the Styrofoam tube holder at your table. Do not open the tube for any reason at this point. WARNING: The reaction mixture contains the deoxyribonucleotides, the primers, and the Taq polymerase. All that is necessary for PCR to start is $Mg^{[++]}$ for the Taq polymerase and a DNA source. Contamination of this tube with an unwanted human DNA will lead to amplification of that DNA. Proceed with the following steps, taking care not to contaminate your sample. Always use clean tips for your pipetter. Never touch the pipette tips to anything except what you are pipetting. Never touch the inside rim of the tube or its lid.

3. Place your reaction mixture tube in the Styrofoam tube holder and add the following:
   a. 15 μl deionized water
   b. 5 μl of your DNA
   c. 10 μl of $MgCl_2$ solution
   Be sure to use a clean pipette tip for each solution. The total volume of the reaction mixture is now 50 μl. *Note:* The $MgCl_2$ solution should be added no more than 20 minutes prior to initiation of the thermocycling.

4. Add one drop of PCR-grade mineral oil to prevent evaporation of solutions during thermocycling. Many automated thermocyclers do not require the use of mineral oil. If this is the case, your instructor will tell you to omit this step.

5. Gently mix the solution by lightly tapping the tube on the table, then place the tube back on ice until your instructor tells you to prepare for thermocycling. Also, remember to return your isolated DNA solution to the ice. At the end of the laboratory period, your instructor may wish to collect these samples and store them for later use.

C. Thermocycling

Thermocycling is the process of shifting the reaction mixture between the denaturation temperature, the primer annealing temperature, and the DNA synthesis temperature. This process is simplified by the use of an automated thermocycler, which is nothing more than a sophisticated, computer-controlled, alternating-temperature chamber. The lack of a thermocycler, however, should not be a major hindrance to performing this investigation; PCR can also be performed by manual temperature cycling. Although somewhat tedious, manual thermocycling illustrates the principle of PCR better than simply putting a sample into a machine.

1.   Automated Thermocycling

If your instructor is using an automated thermocycler, place your tube into the tube holder for the chamber at the front of the laboratory. Most machines will accept 24 samples, so everyone in a standard class should be able to run his or her sample at the same time. The cycler should be set to the following cycling schedule:

   a. 3 minute "hot start" at 94°C
      Cycling

   b. 1 minute 94°C

   c. 1 minute 65°C

   d. 1 minute 72°C
      Repeated for a total of 30 cycles

   e. Finish at 72°C for 10 minutes

   At the end of the cycling schedule, the samples will be returned to 4°C and should be stored at that temperature until electrophoresis can be performed.

2.   Manual Thermocycling (Instructors should carefully review the *Instructor's Manual* several days before performing this procedure.)

Prior to this laboratory period, your instructor set up three 10-inch Corningware "deep-dish" pie dishes containing vegetable oil on magnetic hotplates and adjusted the three temperatures to 94°C, 65°C, and 72°C. Exercise caution; this is hot oil. These three dishes will function in the denaturation, reannealing, and DNA synthesis steps, respectively.

   a. At the front of the laboratory your instructor has constructed a reaction tube holder by stapling two 10 1/4-in. Styrofoam plates together and forming a handle on this "holder" by stapling a couple of thin wires to the plates. In the plate, up to 48 holes the size of the 0.5-ml reaction tube have been punched and numbered. Place your reaction tube in the hole that corresponds to **your number**. Your tube should also be numbered, but this procedure serves as a backup in case your number gets rubbed off with the oil. Keep the plate on ice until all students have "loaded" their reaction tubes.

   b. The manual cycling of these samples will require approximately 2 hours and will require participation of three or four students: one for the actual cycling, one for stopwatch duty and record keeping, and one or two for breaks and trading off. Students in the class should decide by either election, bribery, or coercion who will perform these tasks.

   c. Cycling involves simply transferring the "holder plate" to the appropriate temperature at the appropriate time. Use the stopwatch to keep track of the time in each temperature and use Table 16.1 as your guide and checkoff list for each time and temperature. *Note:* It is important to transfer the reaction tubes quickly between the temperature baths. Cycle the samples as follows:

TABLE 16.1. Cycling Schedule and Check-Off Sheet for Manual Thermocycling

Check off each step as you proceed through the PCR process.

Hot Start 94°C for 3 minutes _____

Cycle Number	94°C	65°C	72°C	Cycle Number	94°C	65°C	72°C
1	1 min _____	1 min _____	1.5 min _____	16	1 min _____	1 min _____	1.5 min _____
2	1 min _____	1 min _____	1.5 min _____	17	1 min _____	1 min _____	1.5 min _____
3	1 min _____	1 min _____	1.5 min _____	18	1 min _____	1 min _____	1.5 min _____
4	1 min _____	1 min _____	1.5 min _____	19	1 min _____	1 min _____	1.5 min _____
5	1 min _____	1 min _____	1.5 min _____	20	1 min _____	1 min _____	1.5 min _____
6	1 min _____	1 min _____	1.5 min _____	21	1 min _____	1 min _____	1.5 min _____
7	1 min _____	1 min _____	1.5 min _____	22	1 min _____	1 min _____	1.5 min _____
8	1 min _____	1 min _____	1.5 min _____	23	1 min _____	1 min _____	1.5 min _____
9	1 min _____	1 min _____	1.5 min _____	24	1 min _____	1 min _____	1.5 min _____
10	1 min _____	1 min _____	1.5 min _____	25	1 min _____	1 min _____	1.5 min _____
11	1 min _____	1 min _____	1.5 min _____	26	1 min _____	1 min _____	1.5 min _____
12	1 min _____	1 min _____	1.5 min _____	27	1 min _____	1 min _____	1.5 min _____
13	1 min _____	1 min _____	1.5 min _____	28	1 min _____	1 min _____	1.5 min _____
14	1 min _____	1 min _____	1.5 min _____	29	1 min _____	1 min _____	1.5 min _____
15	1 min _____	1 min _____	1.5 min _____	30	1 min _____	1 min _____	1.5 min _____

Finish-Up at 72°C for 10 minutes _____

     i. 3 minute "hot start" at 94°C Cycling
    ii. 1 minute at 94°C
   iii. 1 minute at 65°C
   iv. 1.5 minutes at 72°C
       Repeated for a total of 30 cycles
    v. Finish at 72°C for 10 minutes

  d. At the end of the cycling and finish-up time, transfer the plate to ice to cool, and then store the tubes at 4°C until electrophoresis can be performed.

II. AGAROSE GEL ELECTROPHORESIS

Gel electrophoresis is based on the principle of separating molecules by their attraction to an electrical charge. Because DNA contains a negative charge owing to the phosphate groups linking the deoxyribose backbone, it will migrate toward the positive pole when placed in an electrical field. Two major types of gel electrophoresis are used in DNA research, agarose and polyacrylamide (polyacrylamide gel electrophoresis is more difficult to use in a general class setting and hence will not be discussed here). Agarose is a highly purified form of agar and when solidified, will make a network through which DNA must move. The result is that large DNA fragments will migrate more slowly through the agarose, whereas smaller DNA fragments will migrate more rapidly. These different rates allow for the separation of a mixture of DNA fragments by their size (molecular weight). The size of a particular DNA can be used to determine the approximate number of nucleotides in that DNA. *Note:* Other factors such as linear versus circular DNA, supercoiling, and ionic strength of the buffer can also influence the migration of DNA (for a further discussion, see Maniatis et al., 1982; Sambrook et al., 1989).

A.   Preparing the Gel

Although minigels (approximately 7 cm long) can be used for the separation of some PCR fragments, alleles of similar size and some of the MW standards will separate poorly. Consequently, **it is recommended that a larger gel (about 14 cm long)** be used in this investigation. If your instructor decides to use minigels, use one-half the quantities given for grams and milliliters.

1. Weigh 2.0 g agarose in the weighing boat and carefully add the agarose to the 150-ml flask, making sure that none of the agarose touches the sides of the flask.

2. Measure 100 ml electrophoresis buffer in a 100-ml graduated cylinder, and then carefully pour the buffer into the 150-ml flask containing the agarose. This will make a 2.0% agarose gel. Next, add the magnetic stirring bar to the flask and place the flask on the magnetic hotplate. Turn the hot plate to high and the stirrer to a position such that the solution is being completely mixed. Heat the mixture until it starts to boil. The mixture will appear a cloudy white at first, but it will be clear when ready to use.

3. While the agarose is heating, rinse the electrophoresis apparatus with distilled water if it appears dirty, and then dry it with paper towels. Be careful not to damage the thin platinum wire electrodes that are at the bottom of each end of the apparatus.

4. Observing the inside of the electrophoresis apparatus, you will see a central platform, which is open on each end. This area will hold the gel when it is poured. Using the autoclave tape provided, tape the two open ends of the gel platform. Rub your fingernail across the tape where it meets the platform to ensure a tight seal. (Some electrophoresis chambers have the platform at the bottom of the apparatus and have the ends sealed with plastic or metal inserts. If this is the case, your instructor will demonstrate how to set this up.)

5. After taping the ends of the platform, place the Teflon spacer comb on the platform using the holes provided. Place the comb such that the spacers are as far away from one edge of the platform as possible. In addition, make sure the comb teeth do not touch the bottom of the gel apparatus. After the agarose is poured and the gel is solidified, the comb will be removed. The spacers will create wells to which DNA samples can be added. Most combs have six to ten spacers, so that six to ten wells will be present in the gel.

6. Make sure that the electrophoresis apparatus is resting level. When the agarose just starts to boil, remove it from the hotplate using an insulated glove and carefully pour it into the center of the gel platform. Some gel formers require that the agarose be cooled to avoid warping. If this is the case, your instructor will inform you. Pour enough solution so that the agarose is even with the plastic side boundaries of the platform. Make sure there are no strong air drafts that would cause ripples

in the gel, and then allow the gel to solidify. The gel will become opaque when it has solidified. This will require approximately 20 minutes.

7. Once the gel is solidified, remove the tape from the edges of the platform and fill the chamber with electrophoresis buffer so that it completely covers the gel by approximately 1 cm. Allow the gel to rest in the buffer for a minute or two, and then remove the comb by pulling up evenly on both its ends. The buffer should fill the wells when the comb is removed. If a bubble is present in a well, use the Pipetteman to blow buffer into the well, forcing the bubble out. The gels may sit several hours in this state. *Note:* If the comb is removed before the gel is sufficiently solidified, the wells may be damaged, resulting in sample leakage from the wells.

B.   Loading the Gel and Gel Electrophoresis

1. Place a clean 0.5-ml microfuge tube on your Styrofoam tube holder and add 10 µl gel dye. Then, remove your PCR DNA from ice and, after putting a clean tip on your pipette, transfer 20 µl of your PCR DNA to this tube. Remember to wipe off the pipette tip with a Kimwipe to remove any mineral oil. (Return your PCR DNA to ice.) Then, secure the lid of the 0.5-ml tube and gently mix by tapping on the table. If there is an air bubble present, you may need to spin the tube for a couple of seconds in the microfuge.

2. Your instructor will add 20 µl (15 µl DNA and 5 µl gel dye) of DNA standards to the first well of each gel. Adjust your Pipetteman to 30 µl and attach a clean tip. Depress the plunger on the Pipetteman to the first stop, then place the tip to the bottom of the DNA sample and slowly release the plunger. This action will allow 30 µl of the DNA sample to be drawn into the pipette tip. Starting from the well next to the DNA standards and proceeding toward the other end for each member of your group, position the tip of the pipette slightly below the top of a well and slowly depress the applicator button to the first stop position. Because the gel dyes contain compounds denser than the buffer, the DNA sample will flow to the bottom of the well. *Do not depress the applicator button to the blow-out position;* doing so would force air into the well, causing the sample to be ejected.

3. When all members of your group have placed their samples into the wells, put the chamber lid onto the electrophoresis chamber. Attach the black wire to the negative pole and the red wire to the positive pole of the chamber, and then attach the other ends to the respective poles of the power supply. *Note:* DNA is negatively charged; therefore, the positive pole should be at the far end from the sample wells.

4. Set the power supply at 200 V and turn it on for 5 minutes. Then, reduce the voltage to 75 V for the remaining gel running time. The short exposure to the higher voltage serves to quickly stack the DNA against the agarose. *(Warning: You are dealing with an electrical source. Do not touch the buffer or unit while the power is on.)* As the sample migrates through the gel, you will begin to see a blue dye band and a blue-green dye band.

5. Stop the electrophoresis after 90 minutes and replace the buffer (now fairly warm) with fresh chilled buffer. Some electrophoresis chambers get "hotter" more quickly than others, so monitor the buffer temperature and replace the buffer earlier if required. This prevents the gel from over-heating. Run the electrophoresis until the blue dye band is approximately 1–1.5 cm from the bottom of the gel. It will require approximately 3 hours to run this gel.

C.   Staining the Gel

1. Place approximately 200 ml distilled water into a container and add 30 µl ethidium bromide stock solution. Gently swirl the container to mix the stain. *Always wear gloves when working with ethidium bromide because it is a mutagen and potential carcinogen. Your instructor may prefer to stain your gels for you. Note:* Methylene blue staining is not recommended for this investigation.

2. Because it is desirable to flip the gel over for viewing and photographing, it becomes easy to confuse which is the first lane and which is the last lane of DNA. To avoid this potential problem, asymmetry can be introduced into the gel by making a small notch at one side of the gel bottom with a spatula. (*Note:* This is not a concern in this investigation because the DNA markers give asymmetry to the gel.)

3. Carefully slide the gel from the chamber platform to a glass or plastic plate to support the gel. Transfer the gel to the ethidium bromide solution by allowing the gel to slide off the plate gently into the solution. *Note:* Agarose gels can be flimsy and can easily tear, so handle them with care.

4. Place the container on a shaker at low speed for 30 minutes.

5. After the 30-minute staining period, pour the ethidium bromide solution into the waste container provided, being careful not to let the gel slide out of the container.

6. Gently pour approximately 200 ml distilled water over your gel, gently agitate for 1–2 minutes to rinse the gel, and then discard the water into the waste container. Add another 200 ml distilled water and wash the gel for 30 minutes at low speed on the shaker, or simply rock the gel by hand if a shaker is not available. After the 30-minute wash, discard the distilled water into the waste container. (If time permits, a second 30-minute wash in distilled water will produce a gel with less background illumination.)

7. Using a spatula, carefully lift one end of the gel and slide your fingers under it. Lift the gel out of the container and flip it over; then, place it on a piece of Saran Wrap. Since the DNA was at the bottom of the well, flipping the gel over places the DNA closer to the camera for observing and photographing. (Be sure to wear disposable gloves for this procedure.)

8. While it is still on the Saran Wrap, place the gel on the ultraviolet transilluminator which has been positioned behind a plexiglass shield to reduce your exposure to UV light. *Put on plastic safety glasses to protect your eyes further from the UV light, and then turn on the transilluminator.* Ethidium bromide will fluoresce orange anywhere it has intercalated with DNA. If a UV transilluminator is not available, a handheld UV mineral light can serve as a substitute. For viewing a gel with a handheld illuminator, place the Saran-Wrapped gel on the illuminator's glass plate and view directly. You may wish to remove the Saran Wrap from the top of the gel for better viewing. Your instructor will demonstrate how to use the camera and photographic apparatus, if you are photographing the gel. If not photographing the gel, then quickly draw the banding pattern(s) in the space provided in Figure 16.3. (Use the diagram of the DNA size markers as a guide for drawing your bands, and later add the data from other students in your group to the remaining lanes of this figure.)

Answer the following questions about the data you and your class obtained.

1. What proportion of your group is heterozygous and homozygous at the D1S80 locus?_____

    For the entire class? _____

    _____

2. How many different alleles could you detect in your group? _____. Class?

    _____

3. How many individuals can be distinguished by their allelic combination? _____

    _____

4. If an individual possessed two alleles that differed by only one or two repeats, would those

    alleles be detected by agarose gel electrophoresis? _____

    _____

FIGURE 16.3. Photograph for recording PCR electrophoresis data. DNA size standards are shown in lane 1. These DNA size markers range from top to bottom: 10,000 bps, 8000, 6000, 4000, 3000, 2000, 1500, 1400, 1000, 750, 500, 400, 300, and 200. Follow instructions in the text for recording your data.

Look at the DNA size standards shown in Figure 16.3 in answering this question. Would this individual appear as a heterozygote or as a homozygote? _____

_____

5. Why might someone's lane be lacking bands altogether? _____

_____

6. If an individual possessing alleles 18 and 32 were to mate with an individual possessing alleles 16 and 25, determine what allelic combinations their children could have, and, using the DNA markers given in Figure 16.3, indicate what pattern would be observed with agarose gel electrophoresis. Draw these patterns on the edge of Figure 16.3.

III. DETECTION OF DNA POLYMORPHISMS USING RANDOM ANNEALING PRIMERS

The primers designed for the D1S80 locus take advantage of the fact that the DNA sequences flanking either side of this locus are unique and are not found elsewhere in the genome. When given a fairly high-stringency condition for annealing (65°C), these primers will bind in a complementary fashion to only these flanking regions; thus, PCR amplifies only the alleles of this locus. [The higher annealing temperature coupled with the longer length of the primers means that these primers will only base-pair with their exact (or almost exact) complementary sequence in human DNA.] However, if a lower annealing temperature (e.g., 55°C) is used, these primers may also base-pair with similar but not perfectly complementary sequences. This results in additional PCR products and the presence of additional bands in the agarose gel after electrophoresis. Some of these bands may be PCR products of things such as primer-dimer complexes that amplify themselves and hence are artifacts of the PCR process. However, some bands may represent true DNA polymorphisms of similar but different genomic DNA, which are inherited in a Mendelian fashion and can be used to identify different individuals. Thus, there are two major factors that influence the specificity of primers: (1) the length of the primer and (2) the annealing conditions for those primers. The shorter the primer, the greater the number of potential copies of its complementary sequences are in a particular DNA; the lower the annealing temperature, the greater the possible mismatches, i.e., pairing with similar but not exactly the same sequence. These properties can be exploited to detect DNA polymorphisms in organisms by using a technique known as randomly amplified polymorphic DNA (RAPD).

With the RAPD technique, two (one for the left and one for the right) relatively short primers, e.g., 10 base pairs in length, are allowed to anneal to the DNA to be analyzed. Anyplace in the DNA where these two primers anneal within approximately 5000 bps or less of each other can be amplified by PCR. Random probability theory predicts that the number of times this should occur is on the order of 2.5×10^{-9}. Thus, if an organism had a haploid genome size of 4×10^{-9} base pairs, then this annealing should occur approximately 10 times [$(2.5) \times 10^{-9}) \times (4 \times 10^{-9})$], and with PCR approximately 10 bands should be observed. These bands represent distinct sequences in that DNA that should be inherited. If two individuals of the same species differ in one or more of these bands, then this would represent a DNA polymorphism difference between the individuals and thus could be used as an identifying feature (Williams et al., 1990; Welsh and McClelland, 1990).

USE OF RANDOM ANNEALING PRIMERS TO DETECT DNA POLYMORPHISMS[3]

Your instructor may wish to construct an "open-ended" DNA fingerprinting investigation using either human DNA or DNA from some other source. If you are to use human DNA, then you may follow the procedure outlined in Section IA of this investigation. However, you should realize that the cheek cell DNA is contaminated with DNA from other sources and that some of the bands produced with PCR may not be of human origin. If you are using DNA from a different source, you instructor will provide you with the DNA and the background on its origin and isolation.

1. Prepare the PCR DNA reaction tube(s) according to the instructions given in Section IB. Your instructor has included the primers in the reaction mix.
2. Carry out the thermocycling following the steps given in Section IC. The annealing temperature will be different and your instructor will inform you of the temperature that you are to use.

---

[3] Kits containing 10-base oligonucleotide primers can be obtained from Operon Technologies, Inc., 1000 Atlantic Ave., Alameda, CA 94501. Each kit has 20 different primers and material sufficient for approximately 1000 amplifications

3. Prepare and run an electrophoresis following the steps given in Section II; then, record your data in Figure 16.3 using the DNA markers as a guide.

Compare your data with those of the other students in your class and ask such questions as: Can different DNA samples be distinguished? Are some bands present in all or most of the samples? Are identical DNA samples always amplified the same way? If so, what does this suggest? Be sure to discuss your results with your instructor.

REFERENCES

ACUNA, M., H. JORQUERA, L. CIFUENTES, and L. ARMANET. 2002. Frequency of the hypervariable DNA loci D18S849, D3S1744, D12S1090 and D1S80 in a mixed ancestry population of Chilean blood donors. *Genet. Mol. Res.* 1(2):139–146.

BAECHTEL, F.S., J.B. SMERICK, K.W. PRESLEY, and B. BUDOWLE. 1993. Multigenerational amplification of a reference ladder for alleles at locus D1S80. *Journal of Forensic Science* 38(5): 1176–1182.

BARTLETT, J.M.S., and D. STIRLING, eds. 2003. *PCR protocols (methods in molecular biology)*, 2nd ed. Totowa, NJ: Humana Press.

BOWLUS, R.D., and S.C. GRETHER. 1996. A practical polymerase chain reaction laboratory for introductory biology classes. *The American Biology Teacher* 58(3): 172–174.

CANO, R.J., H.N. POINAR, N.J. PIENIAZEK, A. ACRA, and G.O. POINAR. 1993. Amplification and sequencing of DNA from a 120- to 135-million-year-old weevil. *Nature* 363:536–538.

DIEFFENBACH, C.W., and G.S. DVEKSLER, eds. 2003. *PCR primer: a laboratory manual*, 2nd ed. Plainview, NY: Cold Spring Harbor Laboratory Press.

Federation of American Societies for Experimental Biology. 1996. The polymerase chain reaction. *Breakthroughs in Bioscience* (Series). Bethesda, MD: FASEB.

GARRISON, S.J., and C. DE PAMPHILIS. 1994. Polymerase chain reaction for educational settings. *The American Biology Teacher* 56(8): 476–481.

INNIS, M.A., D.H. GELFAND, and J.J. SNINKSY, eds. 1999. *PCR Applications: Protocols for Functional Genomics*. New York: Academic Press.

KLUG, W.S., and M.R. CUMMINGS. 2006. *Concepts of genetics*, 8th ed. Upper Saddle River, NJ: Prentice Hall.

MANIATIS, T., E.F. FRITSCH, and J. SAMBROOK. 1982. *Molecular cloning: a laboratory manual*. Plainview, NY: Cold Spring Harbor Laboratory Press.

MCPHERSON, M.J., B.D. HAMES, and G.R. TAYLOR. 2005. *PCR 2: a practical approach* (Practical Approach Series). New York: Oxford University Press.

MCPHERSON, M.J., P. QUIRKE, and G.R. TAYLOR. 2005. *PCR: a practical approach* (Practical Approach Series). New York: Oxford University Press.

MULLIS, K.B. 1990. The unusual origin of the polymerase chain reaction. *Scientific American* 262(4): 56–65.

NAKAMURA, Y., et al. 1987. Variable number of tandem repeat (VNTR) markers for human gene mapping. *Science* 235:1616–1622.

NAKAMURA, Y., M. CARLSON, K. KRAPCHO, and R. WHITE. 1988. Isolation and mapping of a polymorphic DNA sequence (pMCT 118) on chromosome 1p (D1S80). *Nucleic Acids Research* 16:9364.

NEWTON, C.R. 1995. *PCR: Essential Data*. New York: Wiley-Liss.

NEWTON, C.R., and A. GRAHAM. 1997. *PCR (Introduction to Biotechniques Series)*. New York: Springer-Verlag.

PAABO, S. 1993. Ancient DNA. *Scientific American* 269(5):87–92.

PERKIN-ELMER Corp. AmpliFLP™ D1S80. PCR Amplification Kit User Guide.

PHELPS, T.L., D.G. DEERING, and B. BUCKNER. 1996. Using the polymerase chain reaction in an undergraduate laboratory to produce "DNA fingerprints." The *American Biology Teacher* 58(2):106–110.

SAMBROOK, J. 2005. *Molecular cloning, the condensed protocols: a laboratory manual*. Plainview, NY: Cold Spring Harbor Press.

———, and D.W. RUSSELL. 2001. *Molecular cloning: a laboratory manual* (3- vol. set), 3rd ed. Plainview, NY: Cold Spring Harbor Laboratory Press.

STRACHAN, T., and A.P. READ. 2003. *Human molecular genetics*, 3rd ed. New York: Wiley-Liss.

WELSH, J., and M. MCCLELLAND. 1990. Fingerprinting genomes using PCR with arbitrary primers. *Nucleic Acids Research* 19:303–306.

WILLIAMS, J.G.K., A.R. KUBELIK, K.J. LIVAK, J.A. RAFALSKI, and S.V. TINGEY. 1990. DNA polymorphisms amplified by arbitrary primers are useful as genetic markers. *Nucleic Acids Research* 18:6531–6535.

INVESTIGATION 17

Transformation of *Escherichia coli*

In 1944, Avery, MacLeod, and McCarty published an article (based on research by Griffith) demonstrating that DNA was the agent responsible for conversion (**transformation**) of type IIr nonpathogenic pneumococcal bacteria to a type IIIs pathogenic strain and that, once transformed, the type IIIs were stably inherited. (*Types* refer to an antigenic type and *r* and *s* to the absence or presence of a polysaccharide coat. Note that these two traits are controlled by separate genes.) The work of Avery et al. stimulated interest in DNA and ultimately led to the work of Watson and Crick and formulation of the double-helix model of DNA structure. The fact that a small, isolated DNA molecule could cross a bacterial cell wall and become part of the cell's genetic material contrasted sharply with standard Mendelian genetics but suggested that it might be possible to introduce single genes from one organism into another and, thus, engineer the genotype of the recipient organism. Transformation of DNA forms one of the major bases of molecular genetics.

Today, most transformation procedures utilize **plasmids** (small, circular DNA molecules that have the ability to self-replicate in the cytoplasm of host bacteria), **bacteriophages** (bacterial viruses that can also duplicate in the bacteria), or **cosmids** (a hybrid DNA molecule constructed from plasmid and bacteriophage DNA). Different names are used to refer to the phenomenon of transformation; these include **transduction** and **transfection**. Refer to your textbook for further information. Because DNA molecules can be introduced into bacteria by the process of transformation and can replicate in bacteria, they allow for genetic engineering, or **recombinant DNA technology**. Recombinant DNA molecules can be created when plasmid DNA and a foreign DNA are cut with the same restriction endonuclease to produce complementary single-strand ends. If these two DNAs are mixed, then the foreign DNA can interact with the plasmid DNA via the single-strand ends, thus forming a plasmid and foreign DNA complex, or recombinant DNA molecule (Figure 17.1). This recombinant DNA can then be circularized and used in transformation. Once transformation has occurred, the plasmid containing the foreign DNA can replicate in the bacterium, and, as the bacterium divides, it produces large quantities of the plasmid containing the foreign DNA, which later can be reisolated.

This description of generating a recombinant DNA molecule and carrying out transformation is very simplistic, and many factors are associated with producing the recombinant DNA and transformation. These factors include such things as optimizing the conditions for creating a plasmid–foreign DNA complex and circularizing only recombinant DNA molecules. (Circular DNA transforms at a much higher frequency than does linear DNA.) Other factors, such as the size of the foreign DNA, also influence transformation.

Transformation frequencies for many plasmids are as high as 10^5–10^7 per microgram plasmid DNA. Still, as few as 1 per 10,000 DNA molecules is actually transformed, and many bacteria are not transformed at all. This small proportion of transformed molecules raises the problem of how to distinguish between nontransformed and transformed bacteria. One way that this problem has been alleviated is by incorporating a gene for antibiotic resistance into the plasmid, thereby conveying resistance to the plasmid's host bacterium. Only transformed bacteria will survive when plated onto

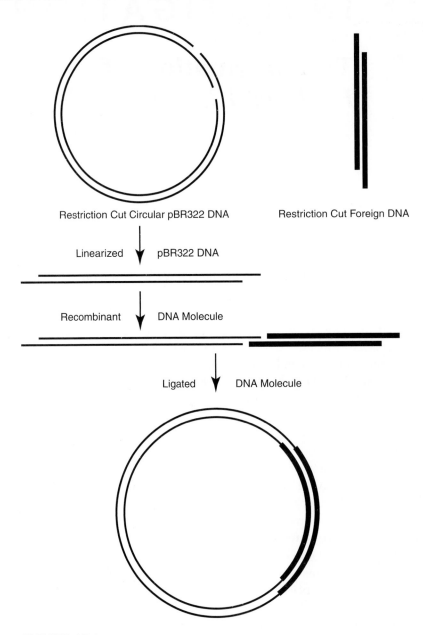

FIGURE 17.1. Insertion of restriction-cut foreign DNA into circular DNA that has been cut with the same restriction endonuclease to produce a recombinant DNA molecule.

antibiotic-containing plates, and essentially all nontransformed bacteria will be killed. The most commonly used antibiotics are tetracycline and ampicillin. The plasmid pBR322 used in this experiment contains genes that confer resistance to these two antibiotics.

High frequencies of transformation are dependent on having bacteria that have been made competent for transformation. Achieving this objective requires altering the bacterial cell wall and membrane such that plasmid DNA can more easily penetrate the bacterium. The most common method of making bacteria competent is by treating the cells with calcium chloride and shocking them with alternating heat and cold.

In today's laboratory work, you will determine the frequency of transformation of *E. coli* using both circular and linear DNA from the plasmid pBR322.

OBJECTIVES OF THE INVESTIGATION

Upon completion of this investigation, you should be able to

1. **define** plasmid, bacteriophage, transformation, and recombinant DNA,

2. **describe** how recombinant DNA molecules are generated and how they produce transformation,

3. **outline** a protocol for determining the frequency of transformation in *E. coli* using both circular and linear DNA from the plasmid pBR322, and

4. **outline** a procedure for determining the relative efficiency of linear vs. circular DNA in producing transformation.

Materials needed for this investigation:

For the class in general:

Escherichia coli (DH5a) made competent for transformation (See appendix at the end of this investigation.)[1]

42°C water bath

circular and linear pBR322 DNA[2] solution (10 ng/μl) (*Note:* The *Instructor's Manual* describes how to prepare the linear DNA.)

37°C incubator with shaker

For each group of students:

3 dilution tubes containing 9.9 ml sterile L-broth, labeled Dil 1, 2, and 3

3 dilution tubes containing 9.0 ml sterile L-broth, labeled Dil 4, 5, and 6

4 sterile agar plates containing tetracycline (12.5 μg/ml), labeled Tet 1–Tet 4

1 sterile agar plate containing tetracycline (12.5 μg/ml) labeled No DNA Tet

3 sterile nutrient agar plates, labeled NA 1–NA 3

1 L-shaped glass rod

1 beaker of 70% ethanol

lighter or alcohol lamp

1 Pipetteman, similar automatic micropipette, or sterile 0.1- and 1.0-ml pipettes

supply of sterile micropipette tips

1 ice tray with ice

3 1.5-ml sterile Eppendorf tubes

1 Eppendorf tube containing frozen competent *E. coli*

1 wax pencil

I.   PROCEDURE FOR TRANSFORMATION OF *E. COLI* USING PLASMID pBR322

1. Prior to class your instructor prepared competent *E. coli* (strain DH5a or a comparable strain that is very susceptible to transformation). Because treated bacteria remain competent for only a day or so, your instructor prepared competent bacteria and then froze and stored them at −80°C using the procedure described in the Appendix to this investigation.

---

[1]   Cultures are available from American Type Culture Collection, 12301 Parklawn Drive, Rockville, MD 20852-1776. *E. coli* cultures and frozen competent *E. coli* cells, as well as plasmids, are also available from Invitrogen 3175 Stanley Rd., Grand Island, NY 14072, and from Carolina Biological Supply Co.

[2]   Available from Sigma Chemical Company, P.O. Box 14508, St. Louis, MO 63178-9916.

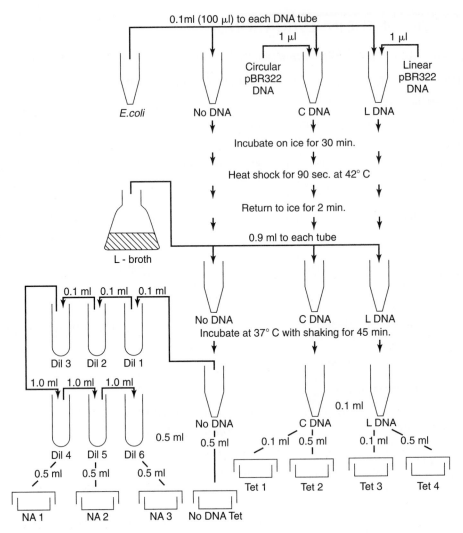

FIGURE 17.2. Protocol for the bacterial transformation experiment. Study this figure in conjunction with the procedure outlined in Section I of the Investigation.

2. Your instructor will provide you with a 1.5-ml Eppendorf tube containing 500 µl frozen competent cells. Place this tube in the ice tray provided, and allow the bacteria to thaw slowly. *Study the flow diagram (Figure 17.2) before proceeding further with the investigation.*

3. While your bacteria are thawing, obtain three 1.5-ml Eppendorf tubes and label them with tape as follows: No DNA, C DNA (for circular), and L DNA (for linear).

4. After your competent bacteria have thawed, gently mix the solution by tilting the tube back and forth. Make sure that you have a homogeneous bacterial suspension.

5. Transfer 100 µl (0.1 ml) of the bacterial suspension to each of the three labeled Eppendorf tubes using the Pipetteman provided. (Your instructor will demonstrate the use of the Pipetteman if you do not know how to use it.)

6. Toward the front of the room, your instructor has placed in an ice tray two tubes of pBR322 DNA at a concentration of 10 ng/1 µl. One tube, labeled C DNA, contains circular pBR322 DNA; the other, labeled L DNA, contains pBR322 DNA that has been restriction cut by a restriction endonuclease and thus is linear. (*Note:* Your instructor may wish to eliminate the linear DNA from this exercise. If so, he or she will inform you of this decision.)

7. Place 1 µl of the C DNA solution into your tube labeled C DNA using a Pipetteman; then, place 1 µl of the L DNA solution into your tube labeled L DNA. *Make sure you use a clean pipette tip*

for each transfer. Recap your tubes and gently mix. *Do not add any DNA to your tube labeled No DNA.*

8. Quickly return all three tubes to your ice tray and allow the bacteria to incubate on ice for 30 minutes.

9. After the 30-minute ice incubation, quickly transfer your tubes to the 42°C water bath for 90 seconds to shock the bacteria.

10. After the 90-second heat shock, return the tubes to ice for 2 minutes. Then, add 0.9 ml (900 μl) L-broth to each of the three tubes and place at 37°C on the shaker for 45 minutes.

11. After the 45-minute incubation, transfer 100 μl (0.1 ml) from the tube labeled C DNA to the tetracycline-containing agar plate Tet 1. Dip the short end of the L-shaped glass rod into a beaker of 70% ethanol; then, remove the rod to a safe distance from the beaker and ignite the alcohol-dipped end with a lighter or alcohol lamp. This operation sterilizes the rod. Cool the rod on the agar before touching the bacteria. Using the glass rod, gently spread the fluid over the surface of the agar plate and then replace the lid. Your instructor may wish to demonstrate this procedure.

12. Next, transfer 500 μl (0.5 ml) from the C DNA tube to a second tetracycline plate (labeled Tet 2), and gently spread the fluid over the plate with the L-shaped rod (sterilized as in step 11).

13. Repeat steps 11 and 12 for the L DNA tube, labeling the tetracycline plates Tet 3 for the 100-μl sample and Tet 4 for the 500-μl sample.

14. Transfer 500 μl (0.5 ml) from the No DNA tube to a tetracycline plate (labeled No DNA Tet).

15. Next, transfer 0.1 ml (100 μl) from the No DNA tube to dilution tube Dil 1 and mix.

16. Transfer 0.1 ml (100 μl) from Dil 1 to dilution tube Dil 2 and mix.

17. Transfer 0.1 ml (100 μl) from Dil 2 to Dil 3 and mix.

18. Transfer 1.0 ml (1000 μl) from Dil 3 to Dil 4 and mix. This process produces a 1×10^{-7} dilution of the bacteria from the "No DNA" tube.

19. Pipet 0.5 ml (500 μl) onto nutrient agar plate NA 1 and spread the fluid with the L-shaped glass rod. (Remember to sterilize and cool the glass rod.)

20. Transfer 1.0 ml from Dil 4 to Dil 5 and mix, producing a 1×10^{-8} dilution. Then, repeat step 18 for NA 2.

21. Transfer 1.0 ml from Dil 5 to Dil 6 and mix, producing a 1×10^{-9} dilution. Repeat step 19 for NA 3.

22. After the fluid has soaked into the agar of the plates, initial each plate with a wax pencil, and place them in an incubator at 37°C with the agar side up for approximately 24 hours.

Note: The No DNA Tet plate serves as a control for insuring that the tetracycline is biologically active and killing most of the nontransformed *E. coli*. Consequently, the number of colonies appearing on the plate after incubation should be zero or very low. An occasional colony will sometimes be observed and represents a spontaneous mutant to tetracycline resistance. (See Investigation 20.) If the number of colonies found on this plate is very small (1 or 2), then the number of spontaneous mutations can be ignored in the calculations for transformation. If the number of colonies is significant (5 or more per plate), then a class average should be determined and this number deducted from the average number of transformed colonies determined from the Tet plates. If the number of colonies is very high, then the antibiotic needs to be replaced.

II. TRANSFORMATION FREQUENCY AND NUMBER OF TRANSFORMANTS PER MICROGRAM OF DNA

The frequency of transformation represents the proportion of bacteria present in the population that took up a pBR322 molecule and thus acquired resistance to the antibiotic tetracycline. The following formulas will allow you to calculate the frequency of transformation for both circular and linear DNA and to compare the efficiency as a transforming agent of circular versus linear pBR322 DNA.

1. Frequency of transformation using circular DNA =

$$\frac{\text{Average no. of colonies on Tet 1 and 2}}{\text{Average no. of colonies on NA 1, 2, and 3}}$$

Remember that you must adjust for the dilution factor (see Table 17.1) before making this calculation as well as those in steps 2 and 3. Enter your calculated value here. _____

2. Frequency of transformation using linear DNA =

$$\frac{\text{Average no. of colonies on Tet 3 and 4}}{\text{Average no. of colonies on NA 1, 2, and 3}}$$

Enter your calculated value here. _____

3. Relative efficiency of linear to circular DNA =

$$\frac{\text{Frequency of linear transformation}}{\text{Frequency of circular transformation}}$$

Enter your calculated value here. _____

4. Transformants per microgram of circular DNA = Average no. of colonies on Tet 1 and 2× (what we call *correction factor* in Table 17.2) × 100. (The reason for multiplying by 100 is that we carried out the transformation using 10 ng DNA.) Enter your calculated value here. _____

5. Transformants per microgram of linear DNA = Average no. of colonies on Tet 3 and 4 (after dilution correction) × 100. Enter your calculated value here. _____

TABLE 17.1.   Data for Transformation Experiment*

Plate Number	Number of Colonies	Correction Factor	Number of Colonies after Dilution Factor Correction	Average
Tet 1	_____	5×	_____	
Tet 2	_____	—	_____	_____
Tet 3	_____	5×	_____	
Tet 4	_____	—	_____	_____
NA 1	_____	$10^7\times$	_____	
NA 2	_____	$10^8\times$	_____	
NA 3	_____	$10^9\times$	_____	_____

* If any of the agar plates contains too many colonies to count, discard that plate in determining the average.

TABLE 17.2. Transformants per Microgram of DNA[*]

Plate Number	Number of Colonies	Correction Factor	Number of Colonies after Dilution Factor Correction	Average
Tet 1	_____	10 ×	_____	
Tet 2	_____	2 ×	_____	_____
Tet 3	_____	10 ×	_____	
Tet 4	_____	2 ×	_____	_____

[*]If any of the agar plates contains too many colonies to count, discard that plate in determining the average.

6. Answer the following questions.
   a. Compare your results for transformation frequency with those of other groups. Were the results from other groups identical? If not, how might you explain the differences? _____

   _____

   _____

   _____

   _____

   b. Briefly discuss your results pertaining to the efficiency of linear versus circular DNA on transformation and what conclusions might be drawn from these results. _____

   _____

   _____

   _____

   _____

APPENDIX

Competent Cell Preparation

Although competent *E. coli* cultures may be purchased, it is much more economical to prepare cultures oneself. To prepare competent cell cultures follow the procedures given below.

Day 1

Inoculate 10 ml sterile L-broth with an *E. coli* strain capable of transformation (e.g., DH5a or other comparable strain) and incubate overnight at 37°C with shaking.

Day 2

1. Transfer the overnight culture to 500 ml L-broth in a 2-liter flask and continue shaking the culture at 37°C. Periodically remove a small sample and determine its optical density (O.D. or absorbancy) at 550 nm.

When the O.D.$_{550}$ = 0.5, there are approximately 5×10^7 cells per milliliter.

2. Chill the culture on ice for 20 minutes. Then, transfer the culture to two chilled and sterile 250-ml centrifuge bottles. It is very important to keep the cells cold at all times.

3. Centrifuge the culture at 5000 rpm in the cold for 5 minutes.

4. Carefully discard the supernatant and gently resuspend each pellet in approximately 50 ml ice-cold competent cell buffer. Balance the bottles.

5. Recentrifuge in the cold at 5000 rpm for 5 minutes. Then, gently discard the supernatant and gently resuspend the cells of each pellet in 125 ml ice-cold competent cell buffer.

6. Place in an ice bath for 1 hour.

7. Recentrifuge in the cold at 5000 rpm for 5 minutes, and then gently pour off the supernatant. Resuspend each pellet in 16 ml cold competent cell buffer. *Cells are very tender at this stage so treat them very gently.* (*Note:* If cells have bound the calcium, they will tend to smear along the sides of the bottles.)

8. Transfer 0.5-ml samples to cold, sterile 1.5-ml Eppendorf tubes; then quick-freeze in liquid nitrogen and store in liquid nitrogen or at −80°C. Competent cells prepared in this way can be stored indefinitely in liquid nitrogen, or for well for over a year at −80°C. If liquid nitrogen is unavailable, competent cells can be frozen on dry ice and stored for a more limited time in a −20°C freezer.

Competent cell buffer per 500 ml

75 ml glycerol	15%
5 ml 1 M Tris, pH 8.0	10 mM
3.675 g CaCl$_2$	50 mM

Add water to bring volume up to 500 ml, filter sterilize, and place on ice until ready to use. Make fresh on the day of the preparation. Nalgene disposable filter sterilization units (0.2 μm) can be obtained from Fisher Scientific.

L-Agar (modified from Maniatis et al., 1982)

10 g NZ amine
5 g NaCl
5 g yeast extract
12.5 g agar
dissolve in 1 liter water
then add 0.70 ml 5 N NaOH
autoclave

L-broth (modified from Maniatis et al., 1982)

10 g NZ amine
5 g NaCl
5 g yeast extract
dissolve in 1 liter water
then add 0.70 ml 5 N NaOH
autoclave

DNA solution (pBR322 DNA in filter-sterilized TE Buffer to give a concentration of 10 ng/μl)

Biological supply companies sell pBR322 DNA in quantities such as 10 μg in a stock solution with concentrations such as 0.1 μg/μl and a total volume of 100 μl. Since 0.1 μg is 100 ng, to create a 10-ng/μl working solution, simply add 1 μl of stock solution to 9 μl of TE buffer. A 0.25-μg/μl stock solution would require diluting 1 μl stock solution with 24 μl TE buffer for a final concentration of 10 ng/μl (250 ng/25μl = 10ng/μl).

TE Buffer pH 7.6

10mM Tris HCl
1 mM EDTA

To make 100 ml of buffer

Add to approximately 90 ml water, 1 ml of 1M Tris-HCl (pH 7.6) and 200 μl 0.5 M EDTA solution, pH 8.0 (Sigma cat.# 03690), adjust to pH 7.6 with HCl, and bring to 100 ml. Filter sterilize.

1M Tris

12.1 g Tris base to 90 ml water, adjust to pH 7.6 with about 6.0 ml HCl, and bring to 100 ml with water.

Tetracycline hydrochloride stock solution (12.5 mg/ml) should be prepared in ethanol-water (50% v/v) and filter sterilized. Tetracycline plates are prepared by autoclaving agar and then allowing the agar to cool to approximately 50°C before adding tetracycline to a final concentration of 12.5 µg/ml. This final concentration of tetracycline is obtained by adding 1 ml of the tetracycline stock solution to 1 liter of melted agar. Tetracycline is light-sensitive, so plates and stock solution should be stored in the dark at 4°C. For information on the stability of antibiotic solutions, see Maniatis et al. (1982). If you decide to use ampicillin instead of tetracycline, use a final concentration of 100 µg/ml.

REFERENCES

AVERY, O.T., C.M. MACLEOD, and M. McCARTY. 1944. Studies on the chemical nature of the substance inducing transformation of pneumococcal types. Induction of transformation by a deoxyribonucleic acid fraction isolated from pneumococcus type III. *Journal of Experimental Medicine* 79:137–158.

BLOOM, M.V., G.A. FREYER, D.A. MICKLOS, S. ZEHL LAUTER. 1996. *Laboratory DNA science: an introduction to recombinant DNA techniques and methods of genome analysis*, New York: Addison-Wesley Publishing Co.

BROCK, T.D. 1990. *The emergence of bacterial genetics*. Plainview, NY: Cold Spring Harbor Laboratory Press.

BROWN, T.A., T. BROWN, and G. SCHOTT, 2006. *Gene Cloning and DNA analysis: an introduction,* 5th ed. Malden, MA: Blackwell Publishing Co.

DALE, J.W., and S.F. PARK. 2004. *Molecular genetics of bacteria*, 4th ed. New York: John Wiley & Sons.

GLOVER, D.M., and B.D. HAMES, eds. 2005. *DNA cloning: a practical approach*, Vol. 1: *Core techniques*, 3rd ed. New York: Oxford University Press.

———. 2005. *DNA cloning: a practical approach*, Vol. 3: *Complex genomes*, 2nd ed. New York: Oxford University Press.

HANAHAN, D. 1987. Mechanisms of DNA transformation, pp. 1177–1183 in F.C. NEIDHARDT, ed. *Escherichia coli and Salmonella typhimurium*, vol. 2. Washington, DC: American Society for Microbiology.

KLUG, W.S., and M.R. CUMMINGS. 2006. *Concepts of genetics*, 8th ed. Upper Saddle River, NJ: Prentice Hall.

LEWIN, B. 2004. *Genes VIII*. Upper Saddle River, NJ: Pearson Prentice Hall.

MANIATIS, T., E.F. FRITSCH, and J. SAMBROOK. 1982. *Molecular cloning: a laboratory manual*. Plainview, NY: Cold Spring Harbor Laboratory Press.

MICKLOS, D.A., and M.V. BLOOM. 1988. DNA transformation of *Escherichia coli*. *Carolina Tips* 51(3):9–11.

MILLER, J.H. 1992. *A short course in bacterial genetics. A laboratory manual and handbook for Escherichia coli and related bacteria*. Plainview, NY: Cold Spring Harbor Laboratory Press.

SAMBROOK, J. 2005. *Molecular cloning, the condensed protocols: a laboratory manual*, NY: Cold Spring Harbor Press.

———, and D.W. RUSSELL. 2001. *Molecular cloning: a laboratory manual*, (3- vol. set), 3rd ed. Plainview, NY: Cold Spring Harbor Laboratory Press.*

SMITH, M.C.M., and E. SOCKETT, eds. 1999. *Genetic methods for diverse prokaryotes (methods in microbiology)*. New York: Academic Press.

SNYDER, L., and W. CHAMPNESS. 2002. *Molecular genetics of bacteria*, 2nd ed. Washington, D.C.: American Society for Microbiology.

---

* Summarized versions of protocols are available online at www.molecularcloning.com for registered book owners.

INVESTIGATION 18

Gene Action: Synthesis of β-Galactosidase in *Escherichia coli*

Gene action is usually studied indirectly as reflected by a modification of a morphological or physiological end product that leads to an altered phenotype. Because genes are known to regulate protein (enzyme) synthesis, investigations can be designed to reveal directly the enzyme produced by the action of a particular gene. In this investigation, the effect of a mutant gene on the activity of the enzyme β-galactosidase in *E. coli* will be studied. Because it provided the first major breakthrough in geneticists' understanding of the control of gene expression (Jacob and Monod, 1961), most genetics textbooks include a discussion of the *lac* operon system in *E. coli.* To prepare for Investigation 18, you are encouraged to review such an account, familiarizing yourself with the biochemical pathway and the vocabulary involved.

I. INTRODUCTION

An *E. coli* culture that is *Lac*$^+$ and inducible has a high β-galactosidase activity when cultured on lactose, a substrate of the enzyme. Lactose (allolactose) acts as an inducer, stimulating the synthesis of the enzyme β-galactosidase. When the same bacteria are grown on another medium, such as nutrient broth, with a different carbon and energy source (e.g., glucose), β-galactosidase is not produced. If organisms of this kind, cultured on nutrient broth, are transferred to a medium with lactose as the only carbon source, exponential growth will occur following a lag period. *Lac*$^-$ *E. coli*, by contrast, grown on nutrient broth and transferred to a medium containing lactose as the only carbon source, will not produce β-galactosidase. Would you expect such *Lac*$^-$ bacteria to grow on a medium in which lac-

tose is the only carbon source? _____

Explain. _____

The detection of the synthesis of β-galactosidase requires bacteria of the proper genotype (*Lac*$^+$), an inducer (lactose), and a substrate (*o*-nitrophenol-β-D-galactoside), which, when split by the enzyme, releases the yellow-colored compound *o*-nitrophenol. Although a qualitative observation of *o*-nitrophenol can be made visually, a quantitative determination of the amount of *o*-nitrophenol released can be obtained spectrophotometrically at 420 nm (the wavelength at which *o*-nitrophenol has maximal absorption of light).

OBJECTIVES OF THE INVESTIGATION

Upon completion of this investigation, you should be able to

1. **outline** a protocol for quantitatively demonstrating the synthesis of the enzyme β-galactosidase in *Escherichia coli*, and

2. **write** a summary statement correlating bacterial growth with β-galactosidase activity, given the spectrophotometric data collected in this experiment.

Materials Needed for this Investigation:

Lac^+ strain[1] of *E. coli* grown with lactose as a carbon source

Lac^+ strain of *E. coli* grown on nutrient broth

Lac^- strain of *E. coli* grown on nutrient broth

concentrated nutrient broth (see recipe on page 210)

minimal medium

    (Recipes for both of these media are given in Section III at the end of this investigation.)

centrifuge and centrifuge tubes

6 250-ml Erlenmeyer flasks

supply of 5-ml pipettes

supply of 1-ml pipettes

37°C incubator or shaker bath

toluene in dropper bottles

0.013 M *o*-nitrophenol-β-D-galactoside in 0.25 M sodium phosphate (neutral pH)

1 M Na_2CO_3 solution

supply of distilled water

Spectronic 20 (or other) spectrophotometer (We recommend that two spectrophotometers be used for each group of students; one set at 420 nm and the other at 600 nm. See Section III for details and for initial adjustment of the instrument.)

spectrophotometer tubes

supply of semilog graph paper (You may photocopy page 306 in this manual.)

II. PROCEDURE

1. Prior to class your instructor standardized the three bacterial cultures by concentrating them by centrifugation at 4000g for 10 minutes, and then by resuspending the bacteria in sterile minimal medium lacking a carbon source. These bacteria are resuspended in enough minimal medium (approximately 100 ml) to give a final optical density (**absorbance**) of 0.10 at a wavelength of 600 nm. Note that optical density and absorbance are equivalent terms. This standardization must be done before class time.

2. Separate each resuspended sample into two equal parts in Erlenmeyer flasks.

3. To one flask of each of the three cultures, add lactose as a carbon source to a final concentration of 1%. To the second flask of each culture, add nutrient broth to a final concentration of 0.8%. (Add 5 ml of concentrated nutrient broth to approximately 45 ml of bacterial solution to achieve a 0.8% final solution.)

---

[1]   One source of Lac^+ and Lac^- *E. coli* strains is the American Type Culture Collection, 12301 Parklawn Drive, Rockville, MD 20852-1776. The Carolina Biological Supply Co. also has Lac^+ and Lac^- *E. coli* strains needed for this investigation.

4. Now, remove a 5-ml sample from each of the six flasks.

   a. Transfer 3 ml to a spectrophotometer tube and read the optical density at 600 nm; record in Table 18.1, column A.

   b. Transfer 1 ml from each flask to a separate test tube, add one drop of toluene to each, and shake vigorously. Now, add 0.20 ml of 0.013 M *o*-nitrophenol-β-D-galactoside in 0.25 M sodium phosphate to serve as the enzyme's substrate. Mix. Incubate these tubes no longer than 20 minutes at 37°C or until a yellow color appears. Stop the reaction by adding 2.70 ml of 1 M Na_2CO_3 and record the time. Determine the optical density of the solution at 420 nm, using the spectrophotometer, and record data in Table 18.1, column B.

   c. Transfer the final 1 ml of each sample to a spectrophotometer tube and add 2.70 ml of 1 M Na_2CO_3 solution and 0.30 ml water. Mix. Each tube should serve as a blank in the enzyme assay. Determine the optical density at 420 nm for each of these tubes, record in Table 18.1 (column C), and subtract from the reading for the corresponding tube containing *o*-nitrophenol-β-D-galactoside.

5. You are now ready to calculate the β-galactosidase activity, expressing the results in enzyme units per sample. One **enzyme unit** may be defined as the amount of enzyme that will produce 1 millimole (mmole) of *o*-nitrophenol per minute. The substrate used in this investigation is *o*-nitrophenol-β-D-galactoside, which is hydrolyzed by the enzyme to produce *o*-nitrophenol, a yellow substance that absorbs light maximally at 420 nm. From a spectrophotometer reading, the amount of *o*-nitrophenol can

TABLE 18.1. Record of Data from Experiment Involving Gene Action

Initial Culture Conditions	Centrifuged and Resuspended in	Spectrophotometric Readings			
		A 3 ml at 600 nm*	B 1 ml Experimental at 420 nm	C 1 ml Control at 420 nm	Difference (B − C)
Lac⁻ strain of *E. coli* grown in nutrient broth	(1) 1% lactose	_____	_____	_____	_____
	(2) 0.8% nutrient broth	_____	_____	_____	_____
Lac⁺ strain of *E. coli* grown with lactose as C-source	(3) 1% lactose	_____	_____	_____	_____
	(4) 0.8% nutrient broth	_____	_____	_____	_____
Lac⁺ strain of *E. coli* grown in nutrient broth	(5) 1% lactose	_____	_____	_____	_____
	(6) 0.8% nutrient broth	_____	_____	_____	_____

* The optical density (absorbance) for these samples initially should be about 0.10. Subsequent readings of the A sample (Table 18.2) will allow one to plot a growth curve for the *E. coli*.

be determined, because one millimole (mmole) of *o*-nitrophenol will yield an optical density of 0.004 at 420 nm. Thus, the optical density reading divided by 0.004 will equal the number of mmoles produced.

a. Now, using this information and the data from Table 18.1, calculate the number of mmoles of *o*-nitrophenol in each of the six samples reported in Table 18.1. Record this information in the spaces that follow.

(1) _____   (4) _____

(2) _____   (5) _____

(3) _____   (6) _____

b. Knowing the time required to produce these mmoles of *o*-nitrophenol, calculate the number of enzyme units produced in each of the six samples and record in the spaces that follow. The number of mmoles produced divided by the total time in minutes equals mmoles/minute, and the number of mmoles/minute divided by 1 mmole/minute equals the number of enzyme units.

(1) _____   (4) _____

(2) _____   (5) _____

(3) _____   (6) _____

c. Why were the *E. coli* cells in step 4b vigorously shaken with toluene? What is the function of

the toluene? _____

_____

_____

_____

6. Now, repeat the procedure of step 4 for each culture every 20 minutes for about 100 minutes or until the culture grown on nutrient broth and transferred to lactose has entered the exponential growth phase. How will you determine when exponential growth has been achieved?

_____

_____

_____

_____

_____

_____

_____

Record all data from these readings in Table 18.2. Repeat the calculations of step 5 for each set of data (40 minutes, 60 minutes, 80 minutes, and 100 minutes) and calculate the number of enzyme units produced in each of the six cultures for each time interval. Record these data in Table 18.3.

TABLE 18.2. Spectrophotometric Readings of the Six Cultures Taken Every 20 Minutes for 100 Minutes

Culture*	20-Minute Readings			40-Minute Readings			60-Minute Readings			80-Minute Readings			100-minute Readings		
	A	B	C	A	B	C	A	B	C	A	B	C	A	B	C
(1)															
(2)															
(3)															
(4)															
(5)															
(6)															

* These six cultures correspond to the six numbered cultures in Table 18.1.

TABLE 18.3. Numbers of Enzyme Units of β-galactosidase Present at Various Time Intervals in Each of the Six Cultures

Culture	Time in Minutes				
	20	40	60	80	100
(1)					
(2)					
(3)					
(4)					
(5)					
(6)					

Note: Data in these columns are calculated using the procedure in step 5b (see text).

For each subculture, plot the logarithm of optical density, representing the growth of *E. coli* and the enzyme activity as functions of time. Clearly identify each curve plotted. These graphs should accompany your laboratory report on this investigation.

Using the spectrophotometric data obtained from cultures growing with lactose as the sole carbon source, compare growth with enzyme activity. Write a brief summary of these comparisons, interpreting them in terms of gene function. _____

_____

_____

_____

_____

_____

_____

_____

III. ADDITIONAL SUGGESTIONS

1. A minimal medium (lacking only a carbon source) for *E. coli* culture may be prepared as follows:
    950 ml distilled water
    50 ml phosphate buffer (*Note:* This buffer is prepared by dissolving 7.26 g KH_2PO4 and 14.34 g K_2HPO4 in 1000 ml distilled water.)
    1.0 g NH_4Cl
    0.10 g Na_2SO_4

    Then to each liter of medium add 2 ml of the following salt solution:
    8.0% $MgCl_2 \cdot 6 H_2O$ (w/v)
    0.10% $CaCl_2 \cdot 2 H_2O$
    0.10% $FeCl_3 \cdot 6 H_2O$

    The pH of the salt solution is adjusted to 2, and the solution is then autoclaved. If the salts precipitate during autoclaving, the solutions may be autoclaved separately and then mixed. The minimal medium described here will support *E. coli* growth when an appropriate carbon source is added. The carbon source should be present in a final concentration of 1%.

2. The medium for growing the *Lac*⁺ *E. coli* will thus consist of the above minimal medium plus 1.0% lactose.

3. Concentrated nutrient broth consists of 8g nutrient broth and 5g NaCl dissolved in 100 ml distilled water.

4. This investigation is most readily conducted if two spectrophotometers are employed—one set at 600 nm and one set at 420 nm. Using two spectrophotometers eliminates the necessity of readjusting the spectrophotometer each time wavelength is changed and thereby decreases the chance for error in obtaining readings. (Initially adjust both spectrophotometers with a spectrophotometer tube containing minimal medium.)

REFERENCES

DALE, J.W., and S.F. PARK. 2004. *Molecular genetics of bacteria,* 4th ed. New York: John Wiley & Sons.

FAIRBANKS, D.J., and W.R. ANDERSEN. 1999. *Genetics: the continuity of life.* Pacific Grove, CA: Brooks/Cole Publishing.

HUDOCK, G.A. 1967. *Experiments in modern genetics.* New York: John Wiley & Sons, pp. 90–96.

JACOB, F., and J. MONOD. 1961. Genetic regulatory mechanisms in the synthesis of proteins. *Journal of Molecular Biology* 3:318–356.

KLUG, W.S., and M.R. CUMMINGS. 2006. *Concepts of genetics,* 8th ed. Upper Saddle River, NJ: Prentice Hall.

LEWIN, B. 2004. *Genes VIII.* Upper Saddle River, NJ: Pearson Prentice Hall.

RUSSELL, P.J. 2006. *iGenetics: a molecular approach.* New York: Benjamin/Cummings Publishing Co.

TAMARIN, R.H. 2001. *Principles of genetics,* 7th ed. New York: McGraw-Hill.

INVESTIGATION 19

Chromatographic Characterization of *Drosophila melanogaster* Mutants

In previous investigations you noted that mutant individuals often depart markedly from normal in their phenotypes. For example, mutant *Drosophila* may have sepia eyes or dumpy wings instead of the normal red eyes and long wings; mutant corn plants may lack chlorophyll and be albino in phenotype. Such altered phenotypes are the result of the altered biochemistry of the mutant individual.

In the early 1940s, George W. Beadle and Edward L. Tatum, working with the mold *Neurospora*, concluded that genes regulate cellular chemistry by controlling the synthesis of organic catalysts called enzymes. Beadle and Tatum suggested what came to be known as the one-gene-one-enzyme hypothesis. Even before the work of Beadle and Tatum, the English physician A.E. Garrod had shown that certain inherited human defects, such as alkaptonuria, were associated with the absence of particular enzymes. However, it has often been difficult to associate enzyme deficiencies with specific gene mutations in complex organisms.

Paper chromatography affords a technique for investigating and characterizing chemical differences between fruit flies having different genetic constitutions. This procedure, first developed for *Drosophila* in 1951 by Ernst Hadorn and Herschel K. Mitchell at the California Institute of Technology, has led to a much better understanding of gene action in *Drosophila*.

According to the procedure devised by Hadorn and Mitchell, fruit flies are crushed along one edge of a rectangular piece of filter paper. The paper is then placed in an appropriate solvent, which, over a period of time, is drawn into the paper by capillary action. As it moves up the paper, the solvent passes through the spots where the fruit flies have been crushed and dissolves substances that are soluble in it. The dissolved substances are then separated from one another by being carried for different distances in the paper, depending on their chemical and physical properties. One group of compounds that can be separated by chromatography and then detected under ultraviolet light is the **pteridines.** Wild-type *Drosophila melanogaster* produces seven pteridines (see Table 19.1).

Flies having mutant eye colors have pteridine patterns that are distinctly different from those of the wild-type flies. Certain normal pteridines may be completely missing, whereas others may be present in abnormally large quantities in flies having mutant genotypes. Hadorn (1962) reported that all eye color mutants studied differ from the wild type and from one another in their pteridine profiles. Paper chromatography thus affords a precise means of characterizing mutant strains of *Drosophila*.

The mutation causing rosy eye color in *Drosophila* has a unique pteridine chromatogram, and affected flies have been shown to lack the enzyme **xanthine dehydrogenase**. Although many of you may find it difficult to distinguish phenotypic differences between wild-type and rosy eye colors, you will find that you can readily detect the differences in their chromatograms.

"Recessive" mutations (e.g., X-linked *w*, white eyes) also affect the pteridine profiles of wild-type heterozygotes. Although a wild-type heterozygote may appear to have normal eye color, chromatography may show it to be not fully normal. This observation has implications for the detection of heterozygosity in human beings who are at risk for being carriers (on the basis of pedigree studies) of

TABLE 19.1. Pteridines in Different Types of *Drosophila melanogaster*

Pigment	Color	Wild Type	Sepia	Brown	Scarlet	White	Other
Isosepiapterin	Yellow	_____	_____	_____	_____	_____	_____
Biopterin	Blue	_____	_____	_____	_____	_____	_____
2-amino-4-hydroxypteridine	Blue	_____	_____	_____	_____	_____	_____
Sepiapterin	Yellow	_____	_____	_____	_____	_____	_____
Xanthopterin	Green-blue	_____	_____	_____	_____	_____	_____
Isoxanthopterin	Violet-blue	_____	_____	_____	_____	_____	_____
Drosopterins	Orange	_____	_____	_____	_____	_____	_____

Note: The pigments are listed in the order in which they will be separated in the chromatogram, with the pigment listed at the bottom of the table being found at the bottom of the chromatogram, nearest to the crushed fly heads.

mutant genes. For example, a test of phenylalanine tolerance can be given to individuals who are suspected of being heterozygous for the "recessive" mutation for phenylketonuria (PKU). Such a test, although imperfect, often can detect heterozygotes, because they have a reduced production of phenylalanine hydroxylase and metabolize phenylalanine less efficiently than homozygous normal individuals. Recent developments in biotechnology make it possible to analyze DNA directly and to do precise detection of heterozygotes for PKU.

In this investigation you will prepare chromatograms of various eye color mutants of *D. melanogaster*. You will then analyze your chromatograms and interpret the results relative to the mutant phenotypes of the flies you observe.

OBJECTIVES OF THE INVESTIGATION

Upon completion of this investigation, you should be able to

1. **outline** a protocol for preparing a chromatogram to demonstrate variations in pteridine pigments in *Drosophila*,

2. **characterize** wild-type and at least four mutant varieties of *Drosophila melanogaster* with respect to seven different pteridine pigments,

3. **discuss** the chromatographic eye pigment patterns in wild-type and mutant *Drosophila* and their relationship to eye color determination, and

4. **discuss** now to calculate R_f values and their potential importance in identifying compounds.

Materials needed for this investigation:

For each pair of students:

dissecting microscope and supplies for handling *Drosophila* (see Investigation 1)

5- × 7-in. rectangle of Whatman no. 1 filter paper

small glass rod

3-lb. coffee can with cover

1000-ml beaker

50 ml solvent: 1:1 mixture of 28% ammonium hydroxide (NH$_4$OH) and *n*-propyl alcohol

2 flies (of the same sex) of each of several different eye color mutants (e.g., wild-type +, sepia [*se*], brown [*bw*], plum [*Pm*], scarlet [*st*], rosy [*ry*], cinnabar [*cn*], vermilion [*v*], eosin [*w^e*], apricot [*w^a*], white [*w*], and a double recessive mutant like cinnabar brown [*cncn,bwbw*] or vermilion brown [*vv,bwbw* and *vY bwbw*])

For the class in general:

ultraviolet light source (a handheld mineral light is sufficient)

stapler

I. PROCEDURE

Compare the following mutant types of *Drosophila* with wild type, using chromatography: sepia,

brown, scarlet, white, _____

Characterize each mutant relative to wild type with respect to the pteridines you have detected by means of chromatography. *Because sex differences exist with respect to the pteridines, the following analyses should be made on flies of the same sex.*

1. Take a 5 × 7-in. rectangle of Whatman no. 1 filter paper and lightly pencil a line 1/2 in. from one of the 7-in. edges and parallel to it. Lightly mark this line with dots at 1-in. intervals. Using dissecting needles, decapitate two etherized wild-type *Drosophila* of the same sex and crush the two heads on the first of these dots using a glass rod. Crush one head at a time and after crushing both heads, wash the glass rod in a solution consisting of equal parts of 28% ammonium hydroxide and *n*-propyl alcohol.

2. Repeat step 1 on the second dot, and then the third dot, etc., using two of each of the mutant flies provided for testing. Remember to space the different types at the 1-in. intervals indicated on the filter paper. Wash the glass rod after each type of fly has been crushed. Try not to touch the filter paper with your fingers. (Why?) Use a pencil to label each spot by letter or number. Record which type of fly was crushed at each site. Allow the spots to air dry for several minutes before proceeding.

3. Now, form the paper into a cylinder about 2 in. in diameter and 5 in. tall by stapling the 5-in. edges together so that they do not overlap. Form this cylinder so that the samples are on the inside of the cylinder. This will allow you to touch the paper without fear of contamination of your samples. Prepare a solvent mixture consisting of equal amounts of 28% ammonium hydroxide and *n*-propyl alcohol.[1] Prior to step 4, place 50 ml of the solvent mixture into a 1000-ml beaker that has been placed inside a large coffee can. Cap the coffee can and let the mixture sit for 5 minutes before using.

4. Insert the filter paper cylinder into the beaker with the crushed fly heads downward. The solvent in the beaker should not touch the spots where the fly heads have been crushed. The filter paper cylinder should not touch the sides of the beaker. Now replace the lid on the coffee can. The chromatogram should be developed in the dark because the pteridines are light-sensitive.

---

[1] The procedure described here has the advantage of permitting the preparation and reading of the chromatogram within a 2-hour laboratory period. Leitenberg and Stokes (1964) have suggested using a solvent mixture of two parts *n*-propanol to one part 1% ammonia, but this solvent requires 16 hours to ascend a 16-in. piece of Whatman no. 1 filter paper. This procedure yields good chromatograms with a well-defined separation of the pteridines, but it is obviously much too slow for completion in a single laboratory period. It must be admitted, however, that the Leitenberg and Stokes procedure produces superior chromatograms.

5. Allow the chromatogram to develop in the closed container for 90 minutes. In this time the solvent front will approach the upper edge of the cylinder and, in doing so, will carry the various pteridine pigments for different distances in the paper. Now remove the paper cylinder from the solvent and allow it to air dry in a standing position for several minutes in the dark. *Caution:* A well-ventilated area will be needed for this purpose.

6. Carefully remove the staples and unroll the dry chromatogram; examine the chromatogram under ultraviolet light in a darkened room or chamber. *Do not look directly at the ultraviolet light.* The names of the seven pteridines found in wild-type flies are listed in Table 19.1. The color each pigment exhibits under ultraviolet light is also indicated. Complete Table 19.1 by checking which of these pigments you can detect in your chromatogram for wild type and for each of the mutant types.

It is now recognized that the various pteridines (listed in Table 19.1) are derived by enzymatic action from 2-amino-4-hydroxypteridine. See Figure 19.1 for a simplified version of these pathways.

II. EXAMINATION OF THE CHROMATOGRAMS

1. Compare your results with those obtained by other students. Did those who used flies of the same

sex as you obtain similar results? _____

If not, suggest possible reasons for the lack of agreement. _____

_____

_____

_____

2. Did those who used flies of the opposite sex obtain different results? _____

_____ If so, in what way did the results differ? _____

_____

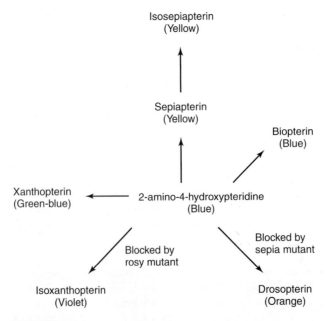

FIGURE 19.1. A simplified version of the biosynthetic pathways for the various pteridines.

3. What might account for sex differences between the flies with respect to the pteridines? _____

_____

4. Can you detect all seven pteridines in the wild-type flies? _____

If not, offer some reasons why you might not be able to. _____

_____

_____

_____

5. Hadorn (1962) reports that sepia-eyed flies completely lack drosopterin, the first of the pteridines.

Does your chromatogram confirm this observation? _____

_____

Note: The drosopterins are red eye pigments. Having observed the phenotype of sepia-eyed flies, you should not be surprised by the lack of red pigment.

6. Does your chromatogram of sepia-eyed flies show any of the pteridines present in amounts in

excess of those found in wild type? _____ Describe your results _____

_____

What pigments were altered? _____

_____

Shown in Figure 19.1 is a simplified version of pteridine biosynthesis. Using Figure 19.1 and your observations of your chromatograms, suggest which pigments are in excess in the sepia-eyed

mutants. _____

Hadorn reports that sepia-eyed flies have excess amounts of 2-amino-4-hydroxypteridine and of

biopterin. Do your observations agree with Hadorn's? _____

If not, suggest some possible reasons for the lack of agreement. _____

_____

Note in Figure 19.1 that a block in the pathway leading to drosopterin results in the sepia phenotype and should increase the concentration of 2-amino-4-hydroxypteridine that is a precursor to several of the other pteridines. The excess of 2-amino-4-hydroxypteridine could potentially increase the amounts of sepiapterin and isosepiapterin. Does this observation change your previous

interpretation? _____

_____

7. What pteridines do you detect in the white-eyed flies? _____

_____

8. What pteridines are present in brown-eyed flies? _____

_____

9. Leitenberg and Stokes (1964) found that adult white- and brown-eye mutants possess no pteridines

   at all! Do your chromatograms support their observations? _____ If not, suggest

   some possible reasons for the lack of agreement. _____

_____

_____

10. Eye color in *D. melanogaster* is the result of the interaction of two groups of pigments, the
    pteridines and brown ommochromes (Strickberger, 1985). What pteridines are present in scarlet-

    eyed flies? _____

_____

_____

11. How do questions 9 and 10 relate to the observed phenotype of the cinnabar, brown double

    mutant flies and the result that this fly stock gave in your chromatogram? _____

_____

12. On the basis of information in questions 7 through 10, predict the phenotypes of the F_1 and F_2 flies
    resulting from a cross between homozygous scarlet-eyed and homozygous brown-eyed flies.

_____

_____

13. In what ways might the various mutant eye color genes tested in this investigation affect the

    adaptability of the flies to their environment? _____

_____

_____

III. CALCULATION OF R_f VALUES

The distance a particular compound is carried in the filter paper is related to the chemical nature of
the compound itself and to the overall distance traveled by the solvent. The distance is characteristic
of the particular compound and is one of the compound's identifying features. The distance is usually
reported in terms of a ratio-to-front (R_f) value, in which

$$R_f = \frac{\text{Distance from the baseline to the center of the pteridine spot}}{\text{Distance from the baseline to the solvent front}}$$

Once empirically established, R_f values allow particular compounds to be identified. R_f values for
a given pteridine are not precisely the same for every chromatogram. Despite this variability, they
become important statistics in characterizing the individual compounds.

TABLE 19.2. Calculation of R_f Values for Pteridine Pigments of a Wild-Type *Drosophila melanogaster*

Pigment	Distance from Baseline to Center of Spot	Distance from Baseline to Solvent Front	R_f
Isosepiapterin			
Biopterin			
2-amino-4-hydroxypteridine			
Sepiapterin			
Xanthopterin			
Isoxanthopterin			
Drosopterins			

You can calculate the R_f value for the various pteridines if (1) you mark (using a pencil) on the chromatogram the solvent front at the time you remove the chromatogram from the developing chamber and (2) you circle the various pigment spots at the time when you examine the chromatogram under ultraviolet light. You can then measure the appropriate distances and calculate R_f values. Now, carry out this procedure and complete Table 19.2 for the pteridines in the wild-type flies. (The solvent that we are using allows for a rapid separation of the pigments but with some smearing of the pigments at their borders. Do the best that you can at locating the center of a pigment band for the calculation of the R_f values.)

REFERENCES

Fairbanks, D.J., and W. Ralph Andersen. 1999. *Genetics: the continuity of life.* Pacific Grove, CA: Brooks/Cole Publishing.

Hadorn, E. 1962. Fractionating the fruit fly. *Scientific American* 206(4):100–110.

Klug, W.S., and M.R. Cummings. 2006. *Concepts of genetics*, 8th ed. Upper Saddle River, NJ: Prentice Hall.

Leitenberg, M., and E.L. Stokes, 1964. *Drosophila melanogaster* chromatography I, II. *Turtox News* 42(9, 10):226–229, 258–260.

Mertens, T.R., and A.S. Bennett. 1969. *Drosophila* chromatography—again. *Turtox News* 47(6):220–221.

Russell, P.J. 1998. *Genetics*, 5th ed. New York: Benjamin/Cummings Publishing Co.

Strickberger, M.W. 1962. *Experiments in genetics with Drosophila* New York: John Wiley & Sons.

———. 1985. *Genetics*, 3rd ed. New York: Macmillan Publishing Company.

Ziegler, I. 1961. Genetic aspects of ommochrome and pterin pigments. *Advances in Genetics* 10:349–403.

INVESTIGATION 20

Bacterial Mutagenesis

In 1953, James Watson and Francis Crick published two papers related to DNA. The first paper described their now-classical model for the structure of DNA. The second discussed the genetic implications of that structure, including the way in which genetic variations (mutations) could occur.

Classically, a **mutation** can be defined as an inheritable alteration of DNA that changes the appearance (phenotype) of an organism. Today, however, we also include as mutations detectable changes in DNA that are inherited but do not produce phenotypic changes in an organism. Such alterations of DNA may be as simple as a single nucleotide substitution or as complex as chromosome rearrangements.

All mutations can be classified as either spontaneous or induced. **Spontaneous mutations** are naturally occurring mutations from a number of causes, such as errors in replication, cosmic radiation, and background radiation. That is, the exact cause of the mutation is unknown. All gene loci have a characteristic spontaneous mutation rate, which may be as high as 10^{-3} or as low as 10^{-8}. If a known agent causes a mutation rate greater than the spontaneous mutation rate, we refer to these mutations as **induced mutations**. Many physical and chemical agents are now known to induce mutations. This investigation will examine both spontaneous and induced mutations to drug resistance in the bacterium *Escherichia coli*.

OBJECTIVES OF THE INVESTIGATION

Upon completion of this investigation, you should be able to

1. **discuss** the differences between spontaneous and induced mutations,

2. **describe** a procedure for determining the spontaneous and induced mutation rates for a particular gene, and

3. **discuss** the induction of mutation by ultraviolet light and the significance of DNA repair mechanisms.

Materials needed for this investigation:

For the class in general:

Escherichia coli culture[1]

nutrient broth culture medium

sterile aqueous 0.1 M magnesium sulfate ($MgSO_4$)

ultraviolet germicidal light (40 W)

For each group of students:

6 dilution tubes containing 9.9 ml sterile nutrient broth, labeled Dil 1 through Dil 3 and Dil 5 through Dil 7

---

[1] Carolina Biological Supply Co. can supply cultures of *E coli*.

2 dilution tubes containing 9.0 ml sterile nutrient broth, labeled Dil 4 and Dil 8

4 nutrient agar plates, labeled NA 1 through NA 4

6 nutrient agar plates containing antibiotic [penicillin (ampicillin) 30 to 100 μg/ml or some other antibiotic], labeled NAA 1 through NAA 6

2 sterile test tubes

1 L-shaped glass rod

1.0- and 0.1-ml sterile pipettes

supply of pipettes (If automatic pipetters are available, they may be used, if using an automatic pipetter then also a supply of sterile tips.)

beaker of 70% ethanol

alcohol lamps or lighter

wax pencil

aluminum foil

I. PROCEDURE FOR DETERMINATION OF SPONTANEOUS MUTATION RATE

In this portion of the investigation, you will determine the spontaneous mutation rate for the gene or genes conveying resistance to an antibiotic. Your instructor will select one of several antibiotics for this investigation (e.g., penicillin, which blocks cell wall growth, or rifampicin, which blocks RNA polymerase activity). Such antibiotics inhibit the growth of any bacterium that does not have a gene conferring resistance to the drug in question. Consequently, the sensitive bacteria eventually die. Bacterial cells possessing genes for resistance are capable of reproducing and forming colonies of bacteria on nutrient agar plates containing the antibiotic.

1. Prior to class, your instructor will grow *E. coli* in 400-ml nutrient broth for 24 hours at 37°C. This procedure produces a concentration of approximately $10^7 – 10^8$ bacteria per milliliter. Just prior to class, the bacteria in this 400 ml of culture will be concentrated by centrifugation at 2000 rpm for 5 minutes and then resuspended in 70 ml of 0.1 M $MgSO_4$ solution. *Study Figure 20.1 before continuing; this figure provides an outline of the entire protocol of the investigation.*

2. Transfer 35 ml of the bacterial $MgSO_4$ suspension to a sterile petri dish for use in Section II of this investigation. Then, add 35 ml nutrient broth to the flask containing the remaining 35 ml of bacterial $MgSO_4$ suspension, swirl to mix, and incubate for 30 minutes. Proceed with step 1 of Section II during this 30-minute incubation period.

3. After 30 minutes of incubation, the instructor will place 2.0 ml of these resuspended bacteria into a sterile test tube for each group of students.

4. Pipet 0.5 ml of the undiluted bacterial culture onto each of the antibiotic nutrient agar plates NAA 1 and NAA 2. Dip the short end of the L-shaped glass rod into 70% ethanol; then, remove the rod to a safe distance and ignite its ethanol-dipped end with a lighter or alcohol lamp. This operation sterilizes the rod. Briefly touch the rod on the surface of the agar to cool it before spreading the bacteria. Now, using the sterile glass rod, gently spread the fluid over the surface of the agar plates. Your instructor may wish to demonstrate this procedure.

5. Transfer 0.1 ml of the undiluted bacterial culture to the Dil 1 tube and mix.

6. Transfer 0.1 ml from Dil 1 to Dil 2 and mix. Then, transfer 0.1 ml from Dil 2 to Dil 3 and mix.

7. Pipet 0.5 ml from Dil 3 to nutrient agar plate NA 1, and gently spread the fluid over the plate with the L-shaped glass rod (sterilized as in step 3).

8. Transfer 1.0 ml from Dil 3 to Dil 4 and mix. Then, pipet 0.5 ml from Dil 4 to plate NA 2 and spread the fluid with the L-shaped glass rod. (Remember to sterilize the glass rod.) *Note:* Plates NAA 1

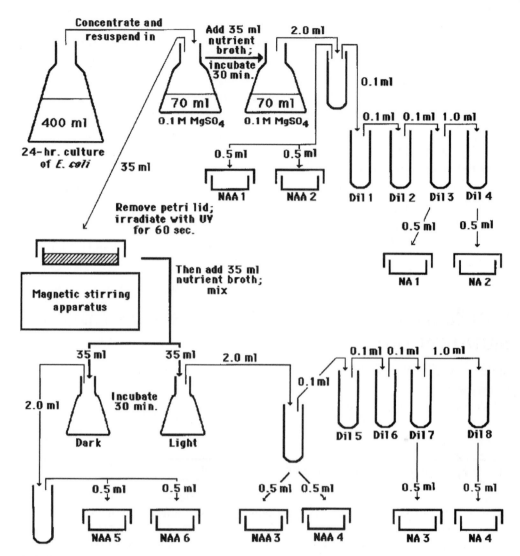

FIGURE 20.1. Protocol for the bacterial mutagenesis experiment. Study this figure in conjunction with the procedure outlined in Sections I and II.

and NAA 2 contain undiluted bacteria. Plate NA 1 has a 10^{-6} dilution of the bacterial culture. Plate NA 2 has a 10^{-7} dilution of the culture.

9. After the fluid has completely soaked into the plates, initial each with a wax pencil; then, place them in an incubator at 37°C for 24 hours with the *agar side up* to prevent the potential smearing of colonies by condensation.

II. ULTRAVIOLET MUTAGENESIS OF *E. COLI*

Does ultraviolet (UV) light induce mutation? The reason for asking this question is twofold: (1) Nucleic acids are known to absorb light in the UV region; (2) thymine dimers as well as other alterations of DNA can result from the absorption of energy released by UV light. In this part of the investigation, you will study the ability of UV light to induce mutation and the effect of visible light on the mutation rate.

1. (To be performed by the instructor.) Add a magnetic stirring bar (sterilized in alcohol and flamed) to the sterile petri dish containing the 35 ml of bacterial MgSO₄ suspension. Place the petri dish on a magnetic stirring apparatus, remove the cover of the petri dish, and irradiate under a germicidal lamp (approximately 40 W) at 30–40 cm for 60 seconds. Add 35 ml of nutrient broth after irradiation,

swirl to mix, and quickly divide the bacterial suspension into two sterile flasks. Label one flask D and completely cover it with aluminum foil; label the other L and leave it exposed to visible light. Incubate both flasks for 30 minutes. At the end of 30 minutes, place 2.0 ml of the L and D bacterial cultures into separate test tubes and distribute one tube of each to each group of students.

2. Pipet 0.5 ml of the culture labeled L onto each of two antibiotic nutrient agar plates (NAA 3 and NAA 4). Spread the fluid over each plate using the sterilized L-shaped glass rod.

3. Make a serial dilution by transferring 0.1 ml of culture L to Dil 5 and mixing. Then, transfer 0.1 ml from Dil 5 to Dil 6 and mix. Transfer 0.1 ml from Dil 6 to Dil 7 and mix. This sequence produces a 10^{-6} dilution of culture L.

4. Pipet 0.5 ml from Dil 7 to nutrient agar plate NA 3 and spread with the sterilized glass rod.

5. Transfer 1.0 ml from Dil 7 to Dil 8 and mix. The final dilution is 10^{-7}.

6. Pipet 0.5 ml from Dil 8 onto nutrient agar plate NA 4 and then spread.

7. Pipet 0.5 ml from culture D onto antibiotic nutrient agar plate NAA 5 and 0.5 ml onto plate NAA 6 and then spread with the sterile L-shaped glass rod.

8. Place your initials on each plate and store the plates with *agar side up* for 24 hours in an incubator at 37°C.

9. After 24 hours, count the number of bacterial colonies on all plates and record the data in Table 20.1.

10. Calculate the spontaneous and induced mutation rates by using the following formulas.

a. Spontaneous mutation frequency =

$$\frac{\text{(Average no. of colonies on NAA 1 and NAA 2)}}{\text{(Average no. of colonies on NA 1 and NA 2)}}$$

TABLE 20.1. Data for Bacterial Mutagenesis Experiment

Plate Number	Number of Colonies	Correction Factor	Number of Colonies after Dilution Factor Correction	Average
NAA 1	_____		_____	
NAA 2	_____		_____	_____
NA 1*	_____	10^6	_____	
NA 2	_____	10^7	_____	_____
NAA 3	_____		_____	
NAA 4	_____		_____	_____
NA 3*	_____	10^6	_____	
NA 4	_____	10^7	_____	_____
NAA 5	_____		_____	
NAA 6	_____		_____	_____

*If either of the nutrient agar plates NA 1 or NA 3 contains too many colonies to be counted, use only NA 2 and NA 4 (times the correction factor) to calculate the mutation frequency.

Remember that you must adjust for the dilution factor (see Table 20.1) before making this calculation.

Enter the value you calculated here._____

b. Induced mutation frequency (visible–light treated) =

$$\frac{\text{(Average no. of colonies on NAA 3 and NAA 4)}}{\text{(Average no. of colonies on NA 3 and NA 4)}}$$

Again, adjust for the dilution factor (Table 20.1).

Enter the value you calculated here._____

c. Induced mutation frequency (dark treated) =

$$\frac{\text{(Average no. of colonies on NAA 5 and NAA 6)}}{\text{(Average no. of colonies on NA 3 and NA 4)}}$$

Enter the value you calculated here._____

11. Answer the following questions._____

a. Does UV light induce mutations?_____

b. If UV light does induce mutations, how much greater is the rate than the spontaneous mutation rate? _____

_____

c. Did you observe any differences in the rate of mutation induced by UV light when your bacterial sample was exposed to visible light or dark conditions? _____

_____

d. How might you explain any differences between the mutation rate of UV-induced mutations in the light-treated versus the dark-treated populations? (You should consult your textbook regarding this question.) _____

_____

_____

_____

_____

REFERENCES

AMES, B.N. 1979. Identifying environmental chemicals causing mutations and cancer. *Science* 204:587–593.

BECKWITH, J., and T.J. SILHAVY, eds. 1992. *The power of bacterial genetics: a literature based course*. Plainview, NY: Cold Spring Harbor Laboratory Press.

BIRGE, E.A. 2000. *Bacterial and bacteriophage genetics*, 4th ed. New York: Springer Publishing.

BROCK, T.D. 1990. *The emergence of bacterial genetics*. Plainview, NY: Cold Spring Harbor Laboratory Press.

DALE, J.W., and S.F. PARK. 2004. *Molecular genetics of bacteria*, 4th ed. New York: John Wiley & Sons.

DEVORET, R. 1979. Bacterial tests for potential carcinogens. *Scientific American* 241(2):40–49.

KLUG, W.S., and M.R. CUMMINGS. 2006. *Concepts of genetics*, 8th ed. Upper Saddle River, NJ: Prentice Hall.

MALOY, R., J. CRONAN, Jr., and D. FREIFELDER. 1994. *Microbial genetics*. Sudbury, MA: Jones and Bartlett Publishers.

MILLER, J.H. 1992. *A short course in bacterial genetics. A laboratory manual and handbook for Escherichia coli and related bacteria*. Plainview, NY: Cold Spring Harbor Laboratory Press.

MULLER, H.J. 1927. Artificial transmutation of the gene. *Science* 66:84–87.

RUSSELL, P.J. 2005. *iGenetics: a Mendelian approach*. New York: Benjamin/Cummings Publishing Co.

STADLER, L.J. 1928. Mutations in barley induced by X-rays and radium. *Science* 68:186–187.

TURN, N.J., and J.E. TREMPY. 2004. *Fundamental bacterial genetics*. Boston: Blackwell Publishing.

WATSON, J.D., and F.H.C. CRICK. 1953. Genetical implications of the structure of deoxyribonucleic acid. *Nature* 171:964–967.

WITKIN, E.M. 1976. Ultraviolet mutagenesis and inducible DNA repair in *Escherichia coli*. *Bacteriological Reviews* 40(4):869–907.

INVESTIGATION 21

Gene Recombination in Phage

Classical investigations by Benzer (1955, 1961) demonstrated gene recombination in bacteriophage. As in studies of recombination in other organisms such as *Drosophila* and maize, it was necessary to obtain two different mutations that could be incorporated into the same organism and could be identified in the progeny of that organism. Benzer accomplished this by infecting *Escherichia coli* cells simultaneously with two different phages, each carrying a genetic marker for plaque morphology. (A **plaque** is a round, clear area on an otherwise opaque culture plate of bacteria, where virulent viruses have lysed the bacterial cells.) The *r*II region of the chromosome of phage T_4 contains two closely linked cistrons, both of which produce rapid lysis resulting in large plaques that can be observed easily on plate cultures of *E. coli*. In Benzer's investigation, each one of these markers was introduced through a different phage.

A genetic "cross" in phage can be accomplished by mixing two different types of mutant phage particles with *E. coli* cells in a broth culture, followed by growing the potentially infected cells in plate cultures. The phage particles become attached (adsorbed) to the host-cell surface, and the phage DNA (but not the protein coat) is injected into the bacterial cell. Within a short time, phage DNA takes over the synthesizing machinery of the bacterial cell, and copies of the phage DNA are then produced. If two different phages have been introduced simultaneously, copies of both types are synthesized in the same cell, and they form a pool of about 100 DNA copies (units). While these DNA copies are in this pool, recombination between the DNAs from the different types of phages can occur. After a short time (usually 10–20 minutes, depending on the phage type), phage DNA units begin to withdraw from the DNA pool and become confined in newly synthesized protein coats. Recombinant phage particles are identified by their ability to grow and produce plaques on culture plates of *E. coli* K.

OBJECTIVES OF THE INVESTIGATION

Upon completion of this investigation, you should be able to

1. **outline** a protocol for detecting wild-type recombinants in the *r*II region of the phage T_4 chromosome by their ability to grow on strains of *E. coli* B and *E. coli* K, and

2. **calculate** the percentage of recombination between *r*II mutants in phage T_4, given data collected in the investigation.

Materials needed for this investigation:

E. coli B, *E. coli* K (lysogenic for phage lambda),[1] two *r* (rapid lysis) mutants of bacteriophage T_4
   Note: Each group of students will need at least 2 ml each of *E. coli* B and *E. coli* K (10^8 per ml) and 1 ml phage mixture (8×10^8 particles per ml).

---

[1] The Carolina Biological Supply Co. catalog lists kits and separate cultures for *E. coli* and phage needed for this investigation.

37°C shaker bath

45°C water bath

37°C incubator

1-ml sterile pipettes

0.10-ml sterile pipettes

pipetters *Note:* Automatic pipetters (e.g., Pipetteman or Finnpipette) should be used if available (100 μl = 0.1 ml). (Sterile pipette tips are also required.)

chloroform in dropping bottle

4 capped tubes, labeled Dil 1, Dil 2, Dil 4, and Dil 5, each containing 9.9 ml sterile nutrient broth

2 capped tubes, labeled Dil 3 and Dil 6, each containing 9 ml sterile nutrient broth

1 125-ml Erlenmeyer flask containing 9.9 ml sterile nutrient broth for a growth tube

12 capped tubes, each containing 2 ml of 0.70% sterile soft agar in distilled water at 45°C; label these tubes SA 1 through SA 12

12 petri plates, each containing sterile Bacto Tryptone medium (1.5% agar) that has been dried in an incubator at 37°C for 24 hours; label four of the petri plates U 1, U 2, U 3, U 4 (unadsorbed phage); label another four B 1, B 2, B 3, B 4 (*E. coli* B for total phage); and label the remaining four K 1, K 2, K 3, K 4 (*E. coli* K for recombinant phage)

1 tube containing 0.10 ml of 2×10^{-3} M potassium cyanide (KCN) solution *(Caution: KCN is extremely poisonous. Take every precaution to avoid accidents. Do not pipet by mouth; use a pipette bulb, a pipette pump, or an automatic pipetter. Thoroughly wash your hands after using this compound).*

I.   PROCEDURE FOR SIMULTANEOUS INFECTION

Note: Before attempting to set up this investigation, read the entire procedure (Parts I–IV) and *study the flow diagram (Figure 21.1) so that you can visualize the sequence of steps you are setting out to accomplish.*

The simultaneous infection of *E. coli* B by two different *r*II mutants is a necessary prequisite for phage recombination. If a time interval should occur between the infection of the two mutants, then exclusion of one or the other type may occur, and, thus, recombination would not be possible. The simultaneous infection of *E. coli* B by the two different *r*II mutants can be ensured by exposing the *E. coli* B cells to two *r*II mutants in the presence of cyanide (KCN). This compound inhibits infection of the bacteria but allows the adsorption of the phage on the bacterial cells. After an appropriate time interval (8 minutes in this investigation), the cyanide is diluted, allowing the adsorbed phage (both *r*II mutants) to infect the bacteria simultaneously.

1. To the tube containing KCN, add 0.50 ml *E. coli* B (10^8 bacteria per ml) and 0.50 ml of phage mixture (8×10^8 particles per ml) with equal parts of two mutant phages (4×10^8 per ml of each mutant).[2] Incubate at 37°C for 8 minutes.

2. Following incubation, transfer 0.10 ml from the KCN culture to the 125-ml flask (growth tube) containing 9.9 ml nutrient broth. Incubate with shaking for 90 minutes at 37°C.

   a. What do you think has taken place while the two mutant phage strains and the *E. coli* B cells were in the KCN tube? _____

_____

_____

---

[2]   Bacteria and phage concentrations must be standardized in advance of class time. For a discussion of the procedures for performing serial dilutions, see pages 14–17 in T.J. Kerr, *Applications in general microbiology: a laboratory manual* (Winston-Salem, NC: Hunter Publishing Company, 1979). Carolina Biological Supply Co. sells individual bacteriophage cultures at concentrations of 10^9 to 10^{11} phage particles per ml and can be diluted to the appropriate concentration.

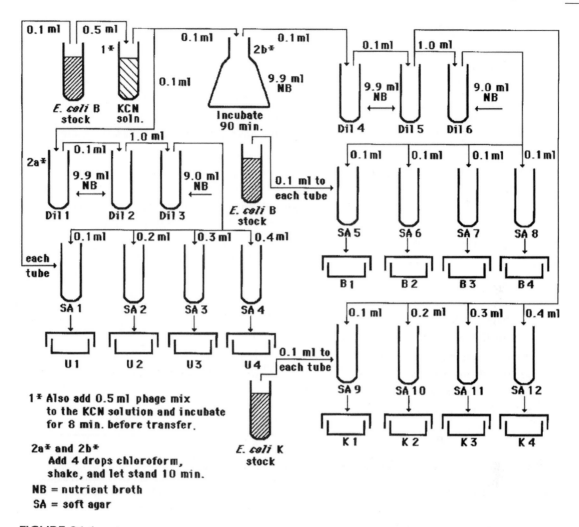

FIGURE 21.1.   Flow Diagram showing the sequence of the procedure outlined in Sections I–IV.

b. How many phage particles relative to each bacterial cell should be present in the KCN tube?

_____

c. What do you expect will happen during the 90 minutes of incubation in the growth tube?

_____

d. Why do you incubate for 90 minutes (one generation)? _____

_____

II.   NUMBER OF UNADSORBED PARTICLES

1. Transfer 0.10 ml from the concentrated KCN culture to the Dil 1 tube and mix. Add 4 drops of chloroform to Dil 1, shake well, and let stand for 10 minutes. The chloroform lyses the bacteria and thus allows the recovery of unadsorbed phage particles, which may then be subjected to further investigation.

2. Transfer 0.10 ml from Dil 1 to Dil 2 and mix.

3. Transfer 1 ml from Dil 2 to Dil 3 and mix. Steps 1, 2, and 3 represent serial dilutions of your initial medium and the phage particles contained therein.

4. Add 0.10 ml *E. coli* B to each of four tubes of soft agar (SA 1, 2, 3, and 4) in a 45°C water bath. To tube SA 1, add 0.10 ml from Dil 3. To tube SA 2, add 0.20 ml from Dil 3. To tube SA 3, add 0.30 ml from Dil 3. To tube SA 4, add 0.40 ml from Dil 3. Mix the contents of tube SA 1 well and spread on the surface of plate U 1. Likewise, mix and spread the contents of tubes SA 2, 3, and 4 on plates U 2, U 3, and U 4, respectively. When the agar has hardened, invert the plates and place them in an incubator at 37°C for 24 hours.

5. After incubation, count the plaques on the U plates. Record these data in Table 21.1 and then multiply by the correction factor indicated in the table. The resulting value represents the number of unadsorbed phage particles present in the growth tube. If the number of particles is small (10% or less of the total particle count as determined in Section III), the unadsorbed phage may be ignored in further calculations.

Why can the unadsorbed phage be ignored (if their number is small) in calculating the percent

of recombination between the *rs*II mutants? _____

_____

_____

_____

_____

TABLE 21.1. Record of Plaque Counts and Calculation of Phage Particles per Milliliter

Plate	Dilution Correction Factors*	Plaque Counts	Corrected Counts (particles/ml)
Unadsorbed Particles			
U 1	10^6	_____	_____
U 2	5×10^5	_____	_____
U 3	3.3×10^5	_____	_____
U 4	2.5×10^5	_____	_____
Total Particles in Growth Tube			
B 1	10^6	_____	_____
B 2	10^6	_____	_____
B 3	10^6	_____	_____
B 4	10^6	_____	_____
Wild-Type Recombinants			
K 1	10^5	_____	_____
K 2	5×10^4	_____	_____
K 3	3.3×10^4	_____	_____
K 4	2.5×10^4	_____	_____

* Multiply these factors times the plaque counts to correct for dilution and thus obtain virus particles per milliliter.

III. TOTAL PARTICLES IN GROWTH TUBE

1. Following 90 minutes (one generation) incubation of the growth tube, add four drops of chloroform to the tube. Shake well and allow to stand for 10 minutes.
2. Transfer 0.10 ml from the growth tube to Dil 4.
3. Transfer 0.10 ml from Dil 4 to Dil 5.
4. Transfer 1 ml from Dil 5 to Dil 6. Note that you have just completed another serial dilution.
5. Add 0.10 ml of the suspension of *E. coli* B to each of four tubes of soft agar (SA 5, 6, 7, and 8) in the 45°C water bath. Transfer 0.10 ml from Dil 6 to each tube. Mix well and spread the contents on plates B 1, B 2, B 3, and B 4. Incubate petri plates in an inverted position at 37°C for 24 hours. Count the number of plaques present on each plate and record these data in Table 21.1.

   a. What is accomplished by this series of steps (i.e., steps 1–5 in Section III)? _____

   _____

   _____

   b. How does this procedure allow one to determine the total number of virus particles present in

   the growth tube? _____

   _____

   _____

   c. What is the total number of phage particles in the growth tube? _____

   _____ Show calculations. _____

   _____

IV. WILD-TYPE RECOMBINANTS

Virus particles that are *r*II mutants can be distinguished from wild-type recombinants on the basis of their effects on the two hosts, *E. coli* B and *E. coli* K. The mutants kill infected *E. coli* K cells but do not lyse them and thus do not liberate progeny phage. Wild-type T₄ phage particles that result from recombination are able to grow, reproduce, and lyse *E. coli* K cells. Therefore, one can determine the proportion of wild-type recombinants by counting the plaques in *E. coli* K cells.

1. To each of four tubes of soft agar (SA 9, 10, 11, and 12) in the 45°C water bath, add 0.10 ml *E. coli* K. From Dil 5 transfer 0.10 ml to tube SA 9, 0.20 ml to tube SA 10, 0.30 ml to tube SA 11, and 0.40 ml to tube SA 12. Mix and spread the contents of these four tubes on the surface of plates K 1, K 2, K 3, and K 4, respectively. After the agar has hardened, invert the plates and incubate at 37°C for 24 hours.
2. Count the number of plaques on the K plates and record these data in Table 21.1.
3. One can determine the percentage of recombination between the two *r*II mutant sites by comparing the numbers of plaques on the B and K plates. If the percentage of unadsorbed phage (see Section II, step 5) is small in Dil 1, the number in the growth tube also will be small and can be ignored. If, however, more than 10% of the phage are unadsorbed (unadsorbed particles in Table 21.1), then this number should be deducted from the total number of particles in the growth tube before

calculating the recombination frequency. Take the indicated correction factors into account when expressing the frequency of recombination. Record all data and make necessary calculations in Table 21.1.

Take an average value for each of the three categories of plates and calculate the percentage of recombination between the two rII mutants, using the following formula.

$$\text{Recombination frequency}^3 = \frac{2 \times \text{Average number of particles/ml on K plates}}{\text{Average number of particles/ml on B plates}} \times 100$$

_____

_____

_____

_____

Why do you multiply the number of wild-type recombinants by 2 when calculating the frequency of recombination? _____

_____

REFERENCES

BENZER, S. 1955. Fine structure of a genetic region in bacteriophage. *Proceedings of the National Academy of Sciences (U.S.A.)* 41:344–354.

———. 1961. On the topography of the genetic fine structure. *Proceedings of the National Academy of Sciences* (U.S.A.) 47:403–415.

BERMAN, D. 1968. The enumeration of bacterial viruses by the plaque technique. *The American Biology Teacher* 30(6):286–287.

BROWN, A.E. 2004. *Benson's microbiological applications: laboratory manual in general microbiology*, 9th ed. Dubuque, IA: McGraw-Hill.

HARTL, D.L., and E.W. JONES. 2004. *Genetics: analysis of genes and genomes*. Sudbury, MA: Jones and Bartlett Publishers.

HAYES, W. 1968. *Genetics of bacteria and their viruses*, 2nd ed. New York: John Wiley & Sons.

HUDOCK, G.A. 1967. *Experiments in modern genetics*. New York: John Wiley & Sons.

KERR. T.J. and B.B. McHALE. 2002. *Applications in general microbiology: a laboratory manual*, 6th ed. Phoenix: Hunter Books.

KLUG, W.S., and M.R. CUMMINGS. 2006. *Concepts of genetics*, 8th ed. Upper Saddle River, NJ: Prentice-Hall.

LEWIN, B. 2004. *Genes VIII*. Upper Saddle River, NJ: Pearson Prentice-Hall.

LEWONTIN, R.C., A.J.F. GRIFFITHS, J.H. MILLER and W.M. GELBART. 2002. *Modern genetic analysis: integrating genes and genomes*, 2nd ed. New York: W.H. Freeman and Company.

PIERCE, B.A. 2005. *Genetics: a conceptual approach*, 2nd ed. New York: W.H. Freeman and Co.

RUSSELL, P.J. 2006. *iGenetics: a molecular approach*. San Francisco: Benjamin Cummings.

SNUSTAD, D.P., and M.J. SIMMONS. 2006. *Principles of genetics*, 4th ed. New York: John Wiley & Sons.

---

[3] Remember that, if fewer than 10% of the phage are unadsorbed, the unadsorbed phage may be ignored in calculating the percentage of recombination between the two rII mutants.

INVESTIGATION 22

Polygenic Inheritance: Fingerprint Ridge Count

Polygenic traits tend to be neglected in the classroom and laboratory despite the fact that in most organisms many significant traits are inherited in this manner. Human examples of polygenic traits often cited in textbooks include skin color in black and white matings, stature, and intelligence as measured by IQ tests. Although these traits do exhibit some of the characteristics associated with polygenic inheritance, they are not illustrated easily with concrete examples in the typical classroom laboratory.

In this investigation, you will explore how the trait of total fingerprint ridge count illustrates the polygenic model of inheritance. Student fingerprint data will be collected and a graphic profile of the class prepared. Experience suggests that most of you are interested in your own fingerprints and those of your peers, so you can expect to find this an interesting investigation to pursue.

I. BACKGROUND

In 1890, Francis Galton suggested fingerprints as a useful tool in personal identification (Galton, 1892; Penrose, 1969). Over the years, the patterns of epidermal ridges and flexion creases on the fingers, toes, palms of the hands, and soles of the feet have become of interest to a variety of specialists. **Dermatoglyphics**, a term coined in 1926 by Harold Cummins, is the study of the epidermal ridges but in practice includes other aspects of hand, finger, and footprints (Penrose, 1969). Fingerprints and other dermatoglyphic data can be obtained from newborns to support clinical diagnoses of chromosome abnormalities such as Down syndrome (see Table 22.1). Although certain dermatoglyphic patterns may be associated with specific chromosome aberrations, be assured that no single fingerprint pattern or ridge count is in itself abnormal.

Although the formation of the epidermal ridge pattern and the total ridge count are polygenic, they are also influenced by environmental factors and thus are more accurately said to be **multifactorial** (Penrose, 1969). The embryology of the epidermal ridges offers clues to the prenatal environmental influence on their pattern of development. Fetal fingertip pads are observable around the sixth week of gestation and reach their maximal size by week 12 or 13, after which they regress, giving rise to elevated dermal ridges (Moore, 1987). The ridges, once formed, are very resistant to later prenatal or postnatal influences, making them an ideal trait for genetic studies as well as for identification of individuals.

II. CLASSIFICATION OF FINGERPRINTS

Fingerprint patterns of dermal ridges can be classified into three major groups: arches, loops, and whorls (see Figure 22.1). The **arch** is the simplest and least frequent pattern. It may be subclassified as "plain" when the ridges rise slightly over the middle of the finger or "tented" when the ridges rise to a point.

The **loop** pattern has a triradius and a core. A **triradius** is a point at which three groups of ridges, coming from three directions, meet at angles of about 120 degrees. The core is essentially a ridge that

TABLE 22.1.   Fingerprint Data That May Be Strongly Suggestive of a Diagnosis for Chromosome Anomalies (Reed, 1981; Penrose, 1969)

Trisomy 21:
Fingers primarily ulnar loops; radial loops on fingers 4 and 5

Trisomy 18:
Underdeveloped epidermal ridges; high frequency of arches (average 7–8; without at least one arch, the diagnosis is suspect); thumbs lacking arches have radial loops; low TRC

Turner syndrome 45,X:
Increased TRC with no increase in whorls

Relationship between average TRC and the number of X and Y chromosomes:

45, X - 165	47, XYY - 103
46, XY - 145	48, XXYY - 88
46, XX - 126	48, XYYY - 83
47, XXY - 114	49, XXXXX - 17 (only 2 individuals examined)

a   b   c

FIGURE 22.1.   Three principal types of fingerprint patterns: (a) arch with no triradius and a ridge count of 0, (b) loop with one triradius and a ridge count of 12, and (c) whorl with two triradii and a ridge count of 15 (the higher of the two possible counts). (Reproduced with permission of the Biological Sciences Curriculum Study from *Basic genetics: a human approach*. Dubuque, IA: Kendall/Hunt Publishing Company, 1983.)

is surrounded by fields of ridges, which turn back on themselves at 180 degrees. Loops can be either radial or ulnar. A finger possesses a **radial loop** if its triradius is on the side of the little finger for the hand in question, and the loop opens toward the thumb. A finger has an **ulnar loop** if its triradius is on the side of the thumb for that hand and the loop opens toward the little finger. The **whorl** pattern has two triradii, with the ridges forming various patterns inside. The frequencies of these fingerprint pattern types in the general population are as follows (Holt, 1968): arch, 5.0%; radial loop, 5.4%; ulnar loop, 63.5%; and whorl, 26.1%.

III. RIDGE COUNT

The focus of this investigation is the polygenic or quantitative trait called the **total ridge count** (TRC), the sum of the ridge counts for all 10 fingers. Holt (1968) found that the average TRC for males is 145 and that for females is 126.

For an arch, the ridge count is 0. The ridge count on a finger with a loop is determined by counting the number of ridges between the triradius and the center or core of the pattern. For a whorl, a ridge count is made from each triradius to the center of the fingerprint, but only the higher of the two possible counts is used (Figure 22.1).

Once all students in the class have prepared their own fingerprints (see Section V) and determined their own TRCs and individual fingerprint patterns, the class can examine how the TRC data support a polygenic model of inheritance.

IV. THE POLYGENIC INHERITANCE MODEL

The inheritance of many significant human behavioral, anatomical, and physiological characteristics is best explained by a polygenic model of transmission. The inheritance of polygenic traits cannot be analyzed by the pedigree method used for single-gene traits, nor by chromosome studies as might be done in the case of suspected chromosomal anomalies. Polygenic traits, in contrast to single-gene traits and chromosome abnormalities, exhibit a wide and continuous range of expression that is measurable. Expression of polygenic traits is often markedly affected by the environment, causing them to be referred to as **multifactorial traits**.

The assumptions underlying the polygenic model of inheritance include the following (Nagle, 1984; Russell, 2006):

- The trait is controlled by many independently assorting gene loci.
- Each gene locus is represented by an active allele that contributes an increment or by an inactive allele that contributes no increment to the phenotype.
- The alleles at each gene locus lack dominance, and each active allele has an effect on phenotype that is small and equal to that of each of the other active alleles affecting the trait.
- Phenotype is determined by the sum of all the active alleles present in the individual.
- Finally, polygenes are not qualitatively different from other genes, they regulate the production of polypeptides, and they segregate and independently assort according to Mendelian principles.

OBJECTIVES OF THE INVESTIGATION

Upon completion of this investigation, you should be able to

1. **construct** a chart of your own fingerprints,
2. **classify** fingerprints into arches, radial and ulnar loops, and whorls,
3. **determine** the total ridge count for a full set of fingerprints,
4. **construct** a histogram using the class data of total ridge counts,
5. **discuss** the characteristics of the polygenic inheritance model and why polygenic traits are more difficult to study than single-gene traits, and
6. **solve** problems concerning TRC by using a four-gene model to explain the inheritance of human fingerprint total ridge counts.

Materials needed for each student for this investigation:

number 2 lead pencil

sheet of plain white paper

roll of 3/4-inch Scotch brand Magic Tape

hand lens, magnifying glass, or dissecting microscope

V. PROCEDURE

The following instructions will provide you with information sufficient to prepare your fingerprints, determine your individual total ridge count, collect class data on TRC, and prepare a histogram of the class data.

1. With a number 2 lead pencil, on a piece of paper shade in a square having sides 3 cm long.

2. Rub one of your fingers in a circular motion on the graphite square, making certain you have covered all of the triradii on the fingerprint. It is important that the sides of your finger be covered with graphite. Now carefully place a piece of Scotch Tape onto your blackened finger so that the tape comes in contact with the entire print. Make certain that you include any triradii on the outer edges of the finger by rolling the finger over the tape in one continuous motion. Peel away the tape and affix it to the appropriate place on your record sheet (Table 22.2).

3. Repeat this process, preparing a print of each of your 10 fingers.

4. Examine each print carefully; if a print is incomplete, prepare a new one. Use a hand lens, magnifying glass, or dissecting microscope to classify the pattern (arch, loop, or whorl) and to determine the ridge count for each print.

5. Record your fingerprint pattern data, total ridge count, and sex in the table on the chalkboard, as directed by the class instructor. Transfer class records from the chalkboard to Table 22.3 and make the calculations indicated in the table.

6. Use the class data to answer the following questions and to construct a histogram (see example in Figure 22.2) in which frequencies are plotted against total ridge count.

Questions. Use the class data recorded on the chalkboard to answer the following questions.

1. What is the average TRC for the class?_____

2. What is the average TRC for the males in the class? _____

   For the females? _____

3. How does your TRC compare to the average for the class? _____

   The average for your sex? _____

4. Is there a difference between male and female average TRCs? _____

   What might account for this difference? _____
   How do the class data compare to the averages published by Holt (1968): 145 for males and 126

   for females?_____

5. In your own words, summarize and describe the histogram you produced from the class data.

   How do the data collected by your class compare to Figure 22.2? _____

   _____

   _____

   _____

TABLE 22.2. Data Sheet for Fingerprints

Right Hand

	Thumb	Second	Third	Fourth	Fifth
Pattern	_____	_____	_____	_____	_____
Ridge count	_____	_____	_____	_____	_____

Total = _____

Place prints
in this space:

Left Hand

	Thumb	Second	Third	Fourth	Fifth
Pattern	_____	_____	_____	_____	_____
Ridge count	_____	_____	_____	_____	_____

Total = _____

Place prints
in this space:

TRC = _____

6. If you had collected TRC data from more people, do you think the histogram for this larger sample

of data would look different from the one you prepared? Explain. _____

_____

_____

_____

_____

TABLE 22.3.   Record of Class Data for Fingerprint Patterns, Total Ridge Count, and Sex of Students

Student	Number of Fingers Having			TRC	Sex
	Loop	Whorl	Arch		
1	_____	_____	_____	_____	_____
2	_____	_____	_____	_____	_____
3	_____	_____	_____	_____	_____
4	_____	_____	_____	_____	_____
5	_____	_____	_____	_____	_____
6	_____	_____	_____	_____	_____
7	_____	_____	_____	_____	_____
8	_____	_____	_____	_____	_____
9	_____	_____	_____	_____	_____
10	_____	_____	_____	_____	_____
11	_____	_____	_____	_____	_____
12	_____	_____	_____	_____	_____
13	_____	_____	_____	_____	_____
14	_____	_____	_____	_____	_____
15	_____	_____	_____	_____	_____
16	_____	_____	_____	_____	_____
17	_____	_____	_____	_____	_____
18	_____	_____	_____	_____	_____
19	_____	_____	_____	_____	_____
20	_____	_____	_____	_____	_____
21	_____	_____	_____	_____	_____
22	_____	_____	_____	_____	_____
23	_____	_____	_____	_____	_____
24	_____	_____	_____	_____	_____
25	_____	_____	_____	_____	_____
Totals	_____	_____	_____	_____	
Percentages of totals	_____	_____	_____		
Mean TRC				_____	
Mean TRC, females				_____	
Mean TRC, males				_____	

Note: A table similar to this can be placed on the chalkboard to collect class data.

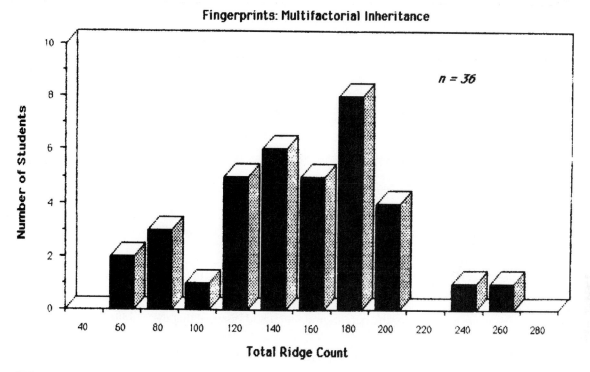

FIGURE 22.2. Total ridge counts for 36 participants in a workshop at Ball State University. (Graph prepared by Richard Menger.)

VI. A SAMPLE OF DATA

The following fingerprint data were collected from 36 individuals participating in a workshop at Ball State University (Figure 22.2). The average TRC for the 19 males in the sample population was 149.2 and that for the females was 129.6. These results compare favorably with those reported by Holt (1968): 145 for males and 126 for females. The frequencies for the different fingerprint patterns for the 36 participants in the workshop also compared favorably with Holt's (1968) data from the general population:

	Workshop participants	General population
Loop	62.2%	68.9%
Whorl	29.7%	26.1%
Arch	8.1%	5.0%
Totals	100.0%	100.0%

VII. EXTEND YOUR UNDERSTANDING WITH ADDITIONAL TRC PROBLEMS

Total fingerprint ridge count exemplifies a polygenic inheritance pattern. Penrose (1969) and others have suggested that a minimum of seven gene loci contribute to TRC, but a four-locus model is hypothesized in the problems that follow. Thus, *AABBCCDD* represents the genotype for maximum ridge count and *aabbccdd* symbolizes the genotype for the minimum ridge count. Assume that each active allele adds 12 ridges to the TRC of the male and 9 to the TRC of the female and that having the genotype *aabbccdd* produces a baseline TRC of 80 for males and 70 for females.

1. Predict the TRC for each of the following individuals.

   Genotype   Male   Female

   AABBCCDD   _____   _____

   AabbccDd   _____   _____

   AaBBCcDD   _____   _____

   aaBbCCDd   _____   _____

2. Write the genotypes of parents who are heterozygous for all four genes.

   _____

   Write the genotype of their child who has the maximum number of active alleles possible.

   _____

   a. What are the TRCs for the parents and their child (assume that the child is a male)?

      Parents: _____

      Child: _____

   b. Calculate the probability that these parents would produce a child with the minimum number of active alleles. Show your calculations.

   _____

3. If an *AaBbCcdd* male mates with an *AaBbCCDD* female,
   a. What is the minimum number of ridge-producing genes possible in one of their children?

   _____

   b. What would be the TRC for this child if it is a male? _____

      A female? _____

   c. If this child is a male, will he have a higher or lower TRC than the parent with the lower ridge

      count? _____
   d. What is the maximum number of ridge-producing genes possible in a child of this couple?

   _____

   e. If this child is a female, will she have a higher or lower TRC than the parent with the higher

      ridge count? _____

      Explain. _____
4. If an *AaBBCcdd* male mates with an *AABbCcDd* female,
   a. What is the minimum number of active alleles possible in a child this couple could produce?

   _____

   b. What would be the probability of producing a child with the minimum number of active alle-

      les? Show your calculations. _____

   _____

   c. What would be the TRC for this child if it were male? _____

      Female? _____

5. How would you expect your TRC to compare with that of your parents? _____

   _____ Your siblings? _____

   Your grandparents?_____

6. In solving problems 1–4, you made some predictions of TRCs based on the genotypes of the individuals involved. Suppose we could measure the TRCs for some people with those genotypes and found the actual values to be different from those predicted by your calculations. How would you explain these discrepancies? _____

   _____

   _____

   _____

   _____

7. Write a paragraph in which you discuss the genetic and environmental components of multifactorial inheritance. _____

   _____

   _____

   _____

   _____

   _____

   _____

   _____

   _____

   _____

REFERENCES

Crawford, M.H., and R. Duggirala. 1992. Digital dermatoglyphic patterns of Eskimo and Amerindian populations: relationships between geographic, dermatoglyphic, genetic and linguistic distances. *Human Biology* 64(5):683–704.

Durham, N.M., and C.C. Plato, eds. 1990. *Trends in dermatoglyphic research.* New York: Springer.

Galton, F. 1892. *Finger prints.* London: Macmillan and Company.

Garruto, R.M., C.C. Plato, and B.A. Schaumann, eds. 1991. *Dermatoglyphics: science in transitions.* Birth Defects Orginal Article Series. New York: Wiley-Liss.

Holt, S.B. 1968. *The genetics of dermal ridges.* Springfield, IL: Charles C. Thomas.

Klug, W.S., and M.R. Cummings. 2006. *Concepts of Genetics,* 8th ed. Upper Saddle River, NJ: Prentice Hall.

Lynch, M., and B. Walsh. 1999. *Genetics and analysis of quantitative traits.* Sunderland, MA: Sinauer Associates Inc.

Mendenhall, G., T. Mertens, and J. Hendrix. 1989. Fingerprint ridge count—A polygenic trait useful in classroom instruction. *The American Biology Teacher* 51(4):203–207.

Moore, L.A. 1987. Dermatoglyphics. *Gene Pool*, January: 1–4. [A Resource Letter for Educators and Students. Dayton, OH: Children's Medical Center.]

Nagle, J.J. 1984. *Heredity and human affairs*, 3rd ed. St. Louis, MO: Times Mirror/Mosby College Publishing.

Nagy, A.S., and M. Pap. 2004. Comparative analysis of dermatoglyphic traits in Hungarian and Gypsy Populations. *Human Biology* 76(3):383–400.

Penrose, L.S. 1969. Dermatoglyphics. *Scientific American* 221(6):72–83.

Reed, T. 1981. Review: Dermatoglyphics in medicine—problems and use in suspected chromosome abnormalities. *American Journal of Medical Genetics* 8:411–429.

Russell, P.J. 2006. *iGenetics a Mendelian approach*. New York: Benjamin/Cummings Publishing Company.

INVESTIGATION 23

Population Genetics: The Hardy-Weinberg Principle

The basic principle of population genetics was formulated in 1908 by an English mathematician, G.H. Hardy, and a German physician, Wilhelm Weinberg.[1] This principle or law is concerned with the frequency of alleles of a gene in a large, randomly interbreeding group of plants or animals. Hardy and Weinberg demonstrated that in such natural populations an equilibrium is reached at which the frequencies of the different alleles of a gene remain constant generation after generation, if no disturbing effects occur, such as those caused by **mutation, selection, random genetic drift, migration,** or **meiotic drive**.

Consider the case in which a gene exists as just two alleles, A and a. Let p equal the frequency of the A allele and q equal the frequency of the a allele. Then, $p + q = 1$. (Why?) It follows that, if random mating occurs, the population should consist of $p^2AA + 2pqAa + q^2aa$ individuals (see Table 23.1).

If a population consists of $p^2AA + 2pqAa + q^2aa$, the next generation, according to the Hardy-Weinberg law, may be expected to consist of exactly the same frequencies of each genotype. The generation-to-generation constancy that exists has been tabulated: The frequencies of different mating combinations are shown in Table 23.2, and the frequencies of the different types of offspring resulting from such matings are summarized in Table 23.3. A study of these tables indicates that if a population is in a Hardy-Weinberg equilibrium, it will stay in that equilibrium generation after generation. Neither the allele nor the genotype frequencies will change as long as random mating continues and there are no disturbing effects caused by mutation, migration, selection, or random genetic drift.

TABLE 23.1. Random Combination of Gametes Results in a Population Consisting of $p^2 AA + 2pq Aa + q^2 aa$

	Sperms	
Ova	$p\ (A)$	$q\ (a)$
$p\ (A)$	$p^2\ (AA)$	$pq\ (Aa)$
$q\ (a)$	$pq\ (Aa)$	$q^2\ (aa)$

---

[1] Some evidence suggests that, as early as 1903, W.E. Castle grasped the basic concept of gene (allele) frequency that is embodied in the Hardy-Weinberg law (see Castle, 1903).

TABLE 23.2. The Determination of the Frequency of Different Matings in a Population Consisting of $p^2AA + 2pqAa + q^2aa$

Females	Males		
	p^2 (AA)	$2pq$ (Aa)	q^2 (aa)
p^2 (AA)	p^4 ($AA \times AA$)	$2p^3 q$ ($AA \times Aa$)	$p^2 q^2$ ($AA \times aa$)
$2pq$ (Aa)	$2p^3 q$ ($AA \times Aa$)	$4p^2 q^2$ ($Aa \times Aa$)	$2pq^3$ ($Aa \times aa$)
q^2 (aa)	$p^2 q^2$ ($AA \times aa$)	$2pq^3$ ($Aa \times aa$)	q^4 ($aa \times aa$)

TABLE 23.3. Frequency of Different Types of Progeny Resulting from Random Matings in a Population in Hardy-Weinberg Equilibrium

Summary of Frequency of Mating	Resulting Progeny from Matings		
	AA	*Aa*	*aa*
p^4 ($AA \times AA$)	p^4		
$4p^3 q$ ($AA \times Aa$)	$2p^3 q$	$2p^3 q$	
$2p^2 q^2$ ($AA \times aa$)		$2p^2 q^2$	
$4p^2 q^2$ ($Aa \times Aa$)	$p^2 q^2$	$2p^2 q^2$	$p^2 q^2$
$4p q^3$ ($Aa \times aa$)		$2pq^3$	$2pq^3$
q^4 ($aa \times aa$)			q^4
Totals	$p^4 + 2p^3 q + p^2 q^2$	$2p^3 q + 4p^2 q^2 + 2pq^3$	$p^2 q^2 + 2pq^3 + q^4$
Totals factored	$p^2 (p^2 + 2pq + q^2)$	$2pq (p^2 + 2pq + q^2)$	$q^2 (p^2 + 2pq + q^2)$
Totals	p^2	$2pq$	q^2

Note: $p^2 + 2pq + q^2 = 1$. Therefore, totals in this row are possible.

OBJECTIVES OF THE INVESTIGATION

Upon completion of this investigation, you should be able to

1. **calculate** the gene (allele) frequencies for a population sample in which each of the genotypes *AA*, *Aa*, and *aa* has a unique phenotype,

2. **determine** whether a population sample in which each of the genotypes *AA*, *Aa*, and *aa* has a unique phenotype represents a population in a Hardy-Weinberg equilibrium,

3. **calculate** the frequencies of the alleles *A* and *a* when dominance is complete (i.e., *AA = Aa* in phenotype), given the frequencies of *A-* and *aa* individuals in a population sample, and

4. **calculate** the frequencies of the I^A, I^B, and i alleles, given a population sample in which the frequencies of individuals having blood types A, B, AB, and O are known.

Materials needed for this investigation:

For each student:

microscope

sterile lancet

"hanging drop" slide with two depressions

filter paper strip

phenylthiocarbamide (PTC) paper strip

For each pair of students:

wax pencil

4 toothpicks

For the class in general:

soap or detergent

5% bleach solution

containers for disposing of utensils and materials that contact blood

paper towels

sterile cotton

70% ethyl alcohol

anti-A and anti-B antisera[2]

I. DETERMINING GENE (ALLELE) FREQUENCIES WHEN CODOMINANCE EXISTS

The Hardy-Weinberg law need not be used to calculate gene frequencies in cases in which alleles exhibit codominance or incomplete dominance. For such cases, allele frequencies can be determined directly from the phenotypes of the individuals involved. Suppose, for example, that you wish to determine the frequencies of the alleles R (red) and r (white) in a population of shorthorn cattle consisting of 63 RR (red), 294 Rr (roan), and 343 rr (white) cattle. Suppose further that you wish to know whether the population is in a Hardy-Weinberg equilibrium. To make these determinations, proceed as follows.

A. Allele Frequencies

In this population of 700 cattle, 63 possess two R alleles and 294 possess one R allele. Thus, among 1400 alleles (2×700) in this population of 700 individuals, there are $(63 \times 2) + 294 = 420$ R alleles. The frequency of $R =$

$$\frac{420}{1400} = \frac{6}{20} = \frac{3}{10} = 0.3$$

Similary, the frequency of $r =$

$$\frac{294 + 343(2)}{1400} = \frac{294 + 686}{1400} = \frac{980}{1400} = \frac{14}{20} = \frac{7}{10} = 0.7$$

B. Hardy-Weinberg Equilibrium

If the population is in a Hardy-Weinberg equilibrium, $(0.3) \times (0.3) = 0.09 = 9\%$ of the population may be expected to be RR (red). Similarly, $2 \times (0.3) \times (0.7) = 0.42 = 42\%$ may be expected to be Rr (roan), while $(0.7) \times (0.7) = 0.49 = 49\%$ may be expected to be rr (white). Applying

---

[2] A more convenient method of typing blood involves using Eldoncards, available through Carolina Biological Supply Co. Although expensive, these cards, which have dried antisera, are more stable and self-contained, eliminating the need for depression slides and transferring of fluids. They also allow typing for Rh blood group.

these percentages to the population of 700 cattle reveals that it is in perfect Hardy-Weinberg equilibrium:

$$9\% \text{ of } 700 = 0.09(700) = 63$$

$$42\% \text{ of } 700 = 0.42(700) = 294$$

$$49\% \text{ of } 700 = 0.49(700) = 343$$

The same technique for determining allele frequencies and then substituting these frequencies in the Hardy-Weinberg genotype distribution might be used to determine whether a human population is in equilibrium for the M and N blood groups. The three possible blood types are type M (genotype MM), type N (genotype $M^N M^N$), and type MN (genotype MM^N).

One study of 1000 Eskimos revealed there to be 835 who were type M, 156 MN, and only 9 who were type N. What are the frequencies of the M and N alleles in this population of 1000? M allele:

_____; N allele: _____.

Does this population of 1000 Eskimos appear to be in a Hardy-Weinberg equilibrium for the M

and N blood groups? _____. Because of random probability, real data and samples will rarely be an exact fit, and observed versus expected results should be compared by the chi-square goodness-of-fit test. (See Investigation 3). Explain whether or not these Eskimo data are in Hardy-Weinberg equilibrium by calculating the expected number of individuals in each of the phenotypes and then calculating chi square to determine how well the sample of 1000 fits the expected

numbers. What is the expected number for each of the phenotypes? Type M _____;

Type MN _____; Type N _____. Now, calculate $\chi^2 =$

_____; the degrees of freedom = _____; the probability that

the deviations are due to chance alone is _____.

Is it reasonable to conclude that this population of 1000 Eskimos is in a Hardy-Weinberg

equilibrium? _____

Is it reasonable, generally, to expect populations to be in a Hardy-Weinberg equilibrium with

respect to traits such as the M and N blood groups? Justify your answer. _____

_____

_____

_____

II.   DETERMINING ALLELE FREQUENCIES USING THE HARDY-WEINBERG LAW

When dominance and recessiveness affect a pair of alleles, it is impossible to detect all three genotypes by their phenotypes and to estimate allele frequencies directly. If, however, you assume the population to be in a Hardy-Weinberg equilibrium, you can estimate allele frequencies, knowing that $q^2 =$ frequency of homozygous recessive individuals. After calculating q by determining the square root of q^2, you can then calculate the frequency (p) of the dominant allele by subtraction, because you know that $p + q = 1$. Knowing the allele frequencies, you can then calculate the frequencies of the homozygous dominant individuals (p^2) and heterozygous dominant individuals ($2pq$).

For example, if you knew that 16% of a certain population was composed of Rh-negative people (rr), you could equate q^2 and 16%.

$$q^2 = 0.16$$

$$q = \sqrt{0.16} = 0.4 = \text{frequency of the allele } r$$

Then, $p = 1 - q = 1 - 0.4 = 0.6 =$ frequency of the allele R. It then follows that the 84% of the population who are Rh-positive can be divided as follows:

$$2pq = 2(0.6)(0.4) = 0.48 = 48\% \text{ of the population expected to be heterozygous, } Rr$$

$$p^2 = (0.6)(0.6) = 0.36 = 36\% \text{ of the population expected to be homozygous, } RR$$

Now, use the Hardy-Weinberg equilibrium to determine the frequencies of the two alleles that regulate the ability to taste the organic compound phenylthiocarbamide (PTC). Estimating the frequencies of these alleles in your genetics class should give you some idea of their frequencies in the population in general.

Studies of individual human families (pedigree studies) have indicated that the ability to taste PTC is due to a dominant allele and that inability to taste the chemical (taste blindness) occurs in homozygous recessive individuals. Thus, there are two phenotypes—"tasters" and "nontasters"—but three genotypes (*TT, Tt,* and *tt*).

A.   Procedure

You will be given a small piece of filter paper that has been impregnated with a dilute solution of PTC. You will use this paper to determine whether or not you are a taster. Before you place the PTC paper on your tongue, taste a piece of untreated filter paper. This will serve as a control so that you will be able to determine readily the difference between the taste of the filter paper and the taste of PTC.

1. Are you a taster of PTC? _____ If so, how did PTC taste? Was it sweet, sour, salty,

    bitter, or some other combination of tastes? _____

    _____

2. You now know your phenotype. Do you know your genotype? _____

    _____

    a. If not, why not? _____

    b. If not, how could you go about finding out what your genotype is? _____

    _____

    _____

3. Record your phenotype in the table provided on the chalkboard. After all of the data are recorded, copy the total class data into the following spaces.

Phenotype	Number	Frequency (decimal fraction)
Tasters	_____	_____
Nontasters	_____	_____
Total	_____	_____

TABLE 23.4. Calculation of Chi-Square for Class Data on Ability to Taste PTC

Phenotype	Observed Number (O)	Expected Number (E)	O − E	$(O - E)^2$	$(O - E)^2/E$
Taster	_____	_____	_____	_____	_____
Nontaster	_____	_____	_____	_____	_____
Totals				$\chi^2 =$	_____

4. Now, calculate the frequency of the recessive allele, t, applying the Hardy-Weinberg equilibrium to the class data. The frequency of t for the class data is _____.

5. The frequency for the T allele is thus _____.
   Using the Hardy-Weinberg equilibrium, you may expect the frequency (p^2) of TT individuals to be _____

   The frequency ($2pq$) of Tt individuals is _____.

B.   Chi-Square Test

Studies of the frequencies of tasters and nontasters have shown that, among the U.S. white population, about 70% can taste PTC, whereas about 30% are taste blind. Other populations show different frequencies of the two phenotypes. For example, some studies show that about 90% of U.S. blacks are tasters and only 10% are nontasters.

   Now, substitute your data in Table 23.4 and calculate chi-square (χ^2) on the assumption that your class is a sample from a population consisting of 70% tasters and 30% nontasters.

1. How many degrees of freedom do you have in interpreting χ^2? _____

2. Does the class seem to be a representative sample of the U.S. white population? _____

   If not, indicate some reasons why the class might not fit this population. _____

   _____

III.   DETERMINING ALLELE FREQUENCIES WHEN THREE ALLELES ARE INVOLVED

In the three preceding examples, the genes involved existed as two alleles (R, r; M, M^N; and T, t). In many situations, however, three or more alleles of a given gene exist in a population. Consider the ABO blood system, with the three alleles I^A, I^B, and i. Let p = frequency of the I^A allele, q = frequency of I^B allele, and r = frequency of the i allele. Then $p + q + r = 1$. (Why?) The frequencies of the different blood groups are tabulated in Table 23.5 for a randomly mating population with the allele frequencies p, q, and r.

   Table 23.5 may be summarized as follows: The population should consist of p^2 (I^AI^A) $+ 2pr$ (I^Ai) $+ q^2$ (I^BI^B) $+ 2$ qr (I^Bi) $+ 2pq$ (I^AI^B) $+ r^2$ (ii). The frequencies of the different phenotypes, then, are as follows:

TABLE 23.5. Random Recombination of Gametes in a Population in Which a Gene Exists as Three Separate Alleles

Ova	Sperms		
	$p\ (I^A)$	$q\ (I^B)$	$r\ (i)$
$p\ (I^A)$	$p^2\ (I^AI^A)$	$pq\ (I^AI^B)$	$pr\ (I^Ai)$
$q\ (I^B)$	$pq\ (I^AI^B)$	$q^2\ (I^BI^B)$	$qr\ (I^Bi)$
$r\ (i)$	$pr\ (I^Ai)$	$qr\ (I^Bi)$	$r^2\ (ii)$

Phenotype	Genotypes	Frequency
A	I^AI^A and I^Ai	$p^2 + 2pr$
B	I^BI^B and I^Bi	$q^2 + 2qr$
AB	I^AI^B	$2pq$
O	ii	r^2

Allele frequencies can now be estimated.

1. Calculation of the frequency of the i allele:

$$r^2 = \text{Frequency of the O phenotype}$$

$$r = \text{Square root of the frequency of the O phenotype}$$

2. Calculation of the frequency of the I^A allele:

$$p^2 + 2pr + r^2 = (p + r)^2 = \text{Frequency of the A phenotype} + \text{Frequency of the O phenotype}$$

$$p + r = \text{Square root of the frequency of the A phenotype} + \text{Frequency of the O phenotype}$$

$$p = \sqrt{p^2 + 2pr + r^2} - r$$

3. Calculation of the frequency of the I^B allele:

This may be obtained, most conveniently, by subtraction, because p and r have already been determined:

$$q = 1 - (p + r)$$

Apply this information to the calculation of p, q, and r for the sample that your class represents. To accomplish this task, type each person's blood and then determine the frequencies of the different blood types.

A. Background

The study of the physiology and genetics of the A, B, O blood groups was initiated by an Austrian, Karl Landsteiner, early in the twentieth century. The results of his studies are found in any beginning genetics textbook and in most general biology textbooks. Landsteiner found that certain **antigens** are associated with the human red blood cell (**erythrocyte**). The antigens are capable of stimulating **antibody** production when they are injected into animals such as rabbits or guinea pigs. The antibodies so produced are then capable of reacting with the specific antigen that caused their production (Table 23.6). Thus, if red blood cells from a type A person were injected into a guinea pig, the guinea pig would produce antibodies to A antigen. Blood serum from the guinea pig would contain these antibodies, which, when mixed with type A human blood, could cause **agglutination** (clumping) of the human erythrocytes.

Table 23.6 shows that the four basic blood groups are controlled by three alleles of a single gene. The I^A and I^B alleles are both **dominant** to the i allele but are **codominant** with respect to each other. Thus $I^A i$ has type A blood and $I^B i$ has type B blood, but $I^A I^B$ possesses both of the antigens and has type AB blood. Notice also that the blood serum of a type A person has antibodies to type B antigen, the blood serum of a type B person has antibodies to type A antigen, the blood serum of a type AB person has neither antibody, and the blood serum of a type O person has antibodies to both A antigen and B antigen.

B. Procedure

In this part of today's investigation you will determine your own blood type with respect to the A and B antigens. *Because of concern for transmitting the viruses causing hepatitis and acquired immune deficiency syndrome (AIDS), great care should be taken in handling human blood.* Tabletops should be washed with a 5% solution of household bleach both before and after this portion of the laboratory investigation. All materials (slides, lancets, swabs, toothpicks, etc.) coming in contact with blood should be placed in special containers for disposal. Hanging drop slides may be autoclaved for reuse. Other materials may be incinerated. Immediately report any accidental blood spills to the instructor. *If your college or university has prescribed procedures on these matters, these procedures should supercede the ones suggested in this investigation. CAUTION is the watchword.*

1. You will be given a glass microscope slide containing two depressions (a "hanging drop" slide). If necessary, clean the slide. Then, using a wax pencil, mark a letter A next to the left depression and a letter B next to the right depression.
2. Carefully wash your hands using the soap or detergent supplied; blot them dry with a *clean* paper towel. Next, swab the tip and side of your left index finger with a sterile cotton swab dipped into 70% ethyl alcohol. Do not dry.

TABLE 23.6. Four Blood Types, Each Associated with a Different Antigen Combination and All Genetically Controlled

Blood Type	Antigen on RBCs	Antibodies Naturally Present in Blood Serum	Possible Genotypes	Approximate Frequencies in U.S. White Population (percent)
A	A	β (anti-B)	$I^A I^A$, $I^A i$	41
B	B	α (anti-A)	$I^B I^B$, $I^B i$	9
AB	A and B	Neither α nor β	$I^A I^B$	3
O	Neither A nor B	Both α and β	ii	47
Total				100

Note: Various studies show somewhat different frequencies for the different blood types. Frequencies given here are a composite of reports given in many introductory genetics and general biology textbooks.

3. Remove a sterile, disposable lancet from its paper cover. While taking care to *touch only the broad end of the lancet*, quickly prick the side of your left index finger. Dispose of the lancet in the container provided.

4. By careful massaging of your lanced finger, squeeze out a drop or two of blood into each depression on the glass slide.

5. Now, add the antisera. To the blood in the depression marked A, add a drop or two of the anti-A (α) serum. To the blood in the depression marked B, add a drop or two of the anti-B (β) serum. Now, use a clean toothpick to mix the blood and serum in the A depression. Use a second toothpick to do the same with the blood and serum in the B depression. Dispose of the toothpicks in the container provided. Why do you use two different toothpicks? _____

_____

The sera you have just used are color coded. What color is the serum containing the α antibodies?

_____

What color is the serum containing the β antibodies? _____

6. Place the slide on a piece of plain white paper and observe. What do you see? _____

_____

Has any noticeable reaction taken place? _____

If so, in which depression on the slide? _____

7. Next, examine the blood and serum mixtures under the low power ($100\times$) of your microscope. Take care not to get any blood on the microscope. Can you make any observations not possible

with the unaided eye? _____

_____

If so, what do you see? _____

_____

The possible reactions are as follows.
a. No agglutination—type O blood
b. Agglutination with anti-A only—type A
c. Agglutination with anti-B only—type B
d. Agglutination with both—type AB

8. What is your blood type, with respect to the A and B antigens? _____
Did you know your blood type before you did this test (e.g., from medical or blood donation

records)? _____

Does your determination agree with what your medical record indicates? _____ If not,

why not? _____

9. When you have determined your blood type, record it on the table provided on the chalkboard. When all of the class data have been recorded on the chalkboard, use these data to complete the following spaces.

Blood type	Number	Frequency (decimal fractions)
A	_____	_____
B	_____	_____
O	_____	_____
AB	_____	_____

C. Allele Frequencies

Now, using this information, show your calculations and determine the following.

1. The frequency of the i allele. _____

_____

2. The frequency of the I^A allele. _____

_____

3. The frequency of the I^B allele. _____

_____

4. What basic assumption are you making about the human population when you calculate allele

frequencies in this fashion? _____

_____

5. Is this a reasonable assumption for the trait of blood types? Justify your answer. _____

_____

D. Calculating Chi-Square

Using your class data, determine whether the class is a random sample of the U.S. white population (see Table 23.6). Make this determination by completing Table 23.7 and calculating χ^2.

1. How do you go about determining the expected numbers in each blood type? _____

_____

2. How many degrees of freedom do you have in interpreting χ^2? _____

_____

3. What is your interpretation of the χ^2 you have just calculated? _____

_____

4. Is the class distribution of blood types comparable to that of the U.S. white population?

_____ If not, why not? _____

_____

_____

_____

_____

TABLE 23.7. Calculation of Chi-Square on Assumption that the Class Is a Representative Sample of the U.S. White Population

Blood Type	O	E	O − E	$(O - E)^2$	$(O - E)^2/E$
A	____	____	____	____	____
B	____	____	____	____	____
AB	____	____	____	____	____
O	____	____	____	____	____
Totals				$\chi^2 =$	_____

REFERENCES

Ayala, F.J. 1982. _Population and evolutionary genetics: a primer._ New York: Addison-Wesley Publishing Company.

Castle, W.E. 1903. The laws of heredity of Galton and Mendel, and some laws governing race improvement by selection. _Proceedings of the American Academy of Arts and Sciences_ 39(8):223–242.

Cavalli-Sforza, L.L. 1974. The genetics of human populations. _Scientific American_ 231(3):80–89.

———. 2001. _Genes, people, and languages._ Berkeley, CA: University of California Press.

———, and W.F. Bodmer. 1999. _The genetics of human populations [unabridged]._ Mineola, NY: Dover Publications.

———, P. Menozzi, and A. Piazza. 1994. _The history and geography of human genes._ Princeton, NJ: Princeton University Press.

Corner, R.C., and T.R. Corner. 1994. Reducing the risk of blood typing: substituting saliva for safety. _The Science Teacher_ 61(3):43–47.

Gillespie, J.H. 2004. _Population genetics: a concise guide,_ 2nd ed. Baltimore: Johns Hopkins University Press.

Halliburton, R. 2003. _Introduction to population genetics._ Upper Saddle River, NJ: Prentice Hall.

Hardy, G. 1908. Mendelian proportions in a mixed population. _Science_ 28:49–50.

Hartl, D.L. 2000. _A primer of population genetics,_ 3rd ed. Sunderland, MA: Sinauer Associates.

———, and A.G. Clark. 1997. _Principles of population genetics,_ 3rd ed. Sunderland, MA: Sinauer Associates.

Hedrick, P.W. 2004. _Genetics of populations,_ 3rd ed. Sudbury, MA: Jones & Bartlett Publishers.

Klug, W.S., and M.R. Cummings. 2006. _Concepts of genetics,_ 8th ed. Upper Saddle River, NJ: Prentice Hall.

Mourant, A.E., A.C. Kopec, and K. Domaniewska-Sobczak. 1976. _The distribution of human blood groups,_ 2nd ed. New York: Oxford University Press.

Sanger, R. 2002. _Human blood groups,_ 2nd ed. Malden MA: Blackwell Publishers.

Spiess, E.B. 1990. _Genes in populations,_ 2nd ed. New York: John Wiley & Sons.

Stern, C. 1962. Wilhelm Weinberg. _Genetics_ 47:1–5.

Note: Additional references related to population genetics are to be found in Investigation 24.

INVESTIGATION 24

Population Genetics: The Effects of Selection and Genetic Drift

In Investigation 23, you learned that the Hardy-Weinberg principle describes a randomly mating population in a state of equilibrium or constancy: The frequencies of alleles *A* and *a* remain constant generation after generation, as do the frequencies of the genotypes *AA*, *Aa*, and *aa*. It was mentioned, however, that the allele frequencies and genotype frequencies remain constant as long as there are no disturbing effects caused by mutation, migration, selection, or random genetic drift.

In this investigation, we will examine the effects of selection and random drift on producing changes in allele frequencies. The effects of **selection** will be studied over a period of weeks in a population of *Drosophila melanogaster*. **Random genetic drift** will be examined in a simulation experiment that will help you to understand how, in a small population, chance alone can lead to the fixation of one allele or another of a given gene.

OBJECTIVES OF THE INVESTIGATION

Upon completion of this investigation, you should be able to

1. **define** the terms *selection* and *random genetic drift*,

2. **design** an experiment to determine the effects of selection on two alleles of a gene in *Drosophila*,

3. **calculate** the changes in gene (allele) frequencies in an experimental population of *Drosophila* and relate these changes to the advantage or disadvantage associated with the alleles in question,

4. **describe** how chance alone can lead to the fixation of different alleles of a gene in different small populations of an organism, and

5. **discuss** the potential role of genetic drift in the evolution of a species.

Materials needed for this investigation:

For each student:

stereo dissecting microscope

etherizer and other equipment for working with *Drosophila*

random numbers table

For the class in general:

Drosophila population cage[1]

Drosophila instant medium

Drosophila stocks selected by the instructor

---

[1] One type of *Drosophila* cage (Drosophila Habitat) is available from Fisher Scientific (cat. no. S17572A), 4500 Turnberry Dr., Hanover Park, IL 60133, and another from Sargent-Welch VWR International (cat. no. WL52465), P.O. Box 5229, Buffalo Grove, IL 60089.

I.   SELECTION OF GENOTYPES AND CHANGE IN ALLELE FREQUENCY

The Hardy-Weinberg principle predicts that allele frequency remains constant in randomly interbreeding populations only if no disturbing effects occur. Allele frequency may change, however, if one genotype is at a disadvantage relative to another genotype. The purpose of this portion of today's investigation is to demonstrate (in *Drosophila*) the change in allele frequency in subsequent generations resulting from selection against a certain genotype.

At the beginning of the course, your instructor chose a trait having two different alleles. An example might be wild-type body color and its recessive allele, ebony (*e*). From each of these homozygous stocks, your instructor isolated 20 virgin females and 20 males. These flies were mixed together and placed in a population cage. The frequency (*q*) of the autosomal recessive allele can be determined by looking at the proportion of ebony flies in the total population. The frequency of ebony flies equals q^2 in the Hardy-Weinberg equation. The frequencies of heterozygotes and homozygous dominants are represented by $2pq$ and p^2, respectively, and can be estimated once you have determined the values of p and q. (For other types of mutants, the frequencies of homozygous dominants and heterozygotes can be determined directly by using a mutant such as X-linked bar eye [*B*] in which heterozygotes have a phenotype distinct from that of either homozygous state.)

Using a + to symbolize the wild-type allele and *e* for ebony, answer the following questions (Included in the *Instructor's Manual* are data collected for six generations from a selection experiment involving wild-type and ebony flies. Your instructor may wish to provide you with these data to assist you in answering these questions).

1. What is the initial frequency of each allele in the parental population of flies?

   + _____ ; *e* _____ After
   1 week, remove the adult flies. Within another week, F$_1$ flies will emerge.

2. What genotype frequency would you expect for +/+ homozygotes?

   _____ +/*e* heterozygotes? _____ *e*/*e* homozygotes?

   _____

   When a large population of F$_1$ flies (200 or more) is present in the cage, remove the adult flies and classify their phenotypes. *Note:* Be careful not to overetherize these flies, because you will need them to produce the next generation. Enter your data in Table 24.1. Calculate the allele and genotype frequencies and enter the values in Table 24.1.

   Replace the used medium in the population cage with fresh medium. Then reintroduce approximately 100 of the *randomly mixed* F$_1$ flies. After 1 week, *remove these adults* and wait for the second generation of flies to emerge. Repeat these steps for each generation, and for as many generations as possible during the duration of the course.

3. Did you observe a change in allele frequency over a period of time? _____

   _____

4. If you observed a decline in the frequency of one of the alleles, what does this mean? _____

   _____

   _____

TABLE 24.1. Data and Calculations of Allele and Genotype Frequencies in a *Drosophila* Population

Generation	Number of Wild-Type Flies	Number of Mutant Flies	Allele Frequency		Frequency of Genotypes		
			p	q	p^2	$2pq$	q^2
0	40	40	___	___			
1	___	___	___	___	___	___	___
2	___	___	___	___	___	___	___
3	___	___	___	___	___	___	___
4	___	___	___	___	___	___	___
5	___	___	___	___	___	___	___
6	___	___	___	___	___	___	___
7	___	___	___	___	___	___	___
8	___	___	___	___	___	___	___

Figure 24.1 graphically depicts the effects of selection on the frequencies of homozygous wild-type (+/+), heterozygous (+/e), and homozygous recessive (e/e) ebony flies over a period of generations, based on the assumption of 80% viability for ebony flies compared to wild-type flies. Note that changes in the frequencies of the alleles + and e over the generations are also shown in the graph.

5. Do the data you collected agree with the predictions of the graph (Figure 24.1)? _____ If not, can you suggest possible reasons why there is a discrepancy? _____

_____

_____

6. Note that although the frequency of homozygous recessive (ebony) flies approaches zero after only a few generations, the frequency of the recessive allele for ebony remains well above zero.

How can this observation be accounted for? _____

_____

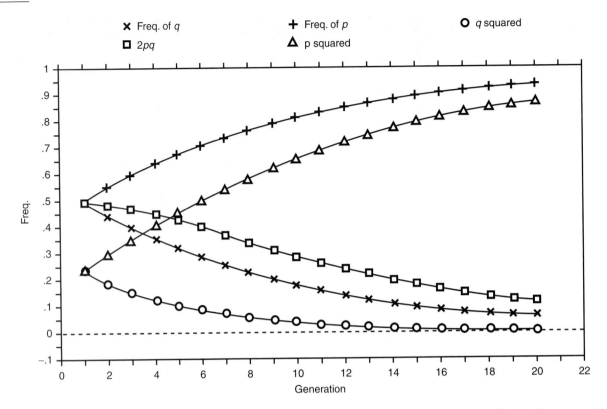

FIGURE 24.1.   Effect of selection on normal and mutant alleles at the ebony locus. See text for details.

II. GENETIC DRIFT AND FIXATION OF AN ALLELE

Genetic drift can be defined as chance fluctuations in gene (allele) frequency in small populations. In large interbreeding populations, the effect of genetic drift can be expected to be small. However, in small "effective breeding populations," genetic drift may be important and may lead to the **fixation** of a specific allele. For example, in a population that possessed two alleles (*A, a*) for a gene, genetic drift could fix the *A* allele so that the entire population would have only the *A* allele and all individuals would be homozygous *AA* (*p* would therefore equal 1). Conversely, genetic drift could fix the *a* allele and all members would be homozygous *aa* with *q* = 1. The term **effective breeding population** is used because the real breeding population may be much smaller than the actual population. For example, with humans, a large segment of the population is nonbreeding because of age or even artificial birth control. In addition, in many populations individuals may be isolated into smaller breeding units even though gene flow may occasionally occur between these units. For example, gorillas are organized into troupes consisting of anywhere between 9 and 12 breeding males and females. Thus, a troupe comprises a small effective breeding population, even though occasional interbreeding between troupes may occur.

The effects of genetic drift with random interbreeding can be simulated by using a random numbers table based on 10 numbers, 0–9. Assume, for example, that there is a gene with two alleles (*A* and *a*) that are present initially in equal proportions (*p* = 0.5 and *q* = 0.5) and randomly distributed among the possible genotypes within a population of five individuals. This means that there are a total of 10 alleles (5 *A* alleles and 5 *a* alleles). We can use 10 numbers from 0 to 9, and assign numbers 0, 1, 2, 3, and 4 to represent allele *A* and numbers 5, 6, 7, 8, and 9 to represent allele *a*. If we now choose at random 10 numbers from a random numbers table—for example, 0076361995—this sequence can represent a random breeding and mixing of alleles. Because allele *A* is represented by numbers 0 through 4 and allele *a* by numbers 5 through 9, the alleles present in the genotypes of this first generation are *AAaaAaAaaa*. Thus, allele *A* accounts for 40% of the total alleles (4 of a total of 10 alleles) in the generation produced. Because there are four *A* alleles in this generation, the numbers

TABLE 24.2. Simulation Experiment of Genetic Drift Using Random Numbers (data generated by R.L. Hammersmith)

Generation	Transforming Random Numbers into Alleles	Random Number	Resulting Alleles	Frequency of *A* Alleles (percent)
0	—	—	*AAAAAaaaaa*	50
1	0, 1, 2, 3, 4 = *A*; 5, 6, 7, 8, 9 = *a*	0076361995	*AAaaAaAaaa*	40
2	0, 1, 2, 3 = *A*; 4, 5, 6, 7, 8, 9 = *a*	0010721254	*AAAAaAAAaa*	70
3	0, 1, 2, 3, 4, 5, 6 = *A*; 7, 8, 9 = *a*	9393330184	*aAaAAAAAaA*	70
4	0, 1, 2, 3, 4, 5, 6 = *A*; 7, 8, 9 = *a*	1679333749	*AAaaAAAaAa*	60
5	0, 1, 2, 3, 4, 5 = *A*; 6, 7, 8, 9 = *a*	9623916348	*aaAAaAaAAa*	50
6	0, 1, 2, 3, 4 = *A*; 5, 6, 7, 8, 9 = *a*	1122498507	*AAAAAaaaAa*	60
7	0, 1, 2, 3, 4, 5 = *A*; 6, 7, 8, 9 = *a*	0496056062	*AAaaAAaAaA*	60
8	0, 1, 2, 3, 4, 5 = *A*; 6, 7, 8, 9 = *a*	0381078692	*AAaAAaaaaA*	50
9	0, 1, 2, 3, 4 = *A*; 5, 6, 7, 8, 9 = *a*	2685322676	*AaaaAAAaaa*	40
10	0, 1, 2, 3 = *A*; 4, 5, 6, 7, 8, 9 = *a*	5125959529	*aAAaaaaaAa*	30
11	0, 1, 2 = *A*; 3, 4, 5, 6, 7, 8, 9 = *a*	6999923877	*aaaaaAaaaa*	10
12	0 = *A*; 1, 2, 3, 4, 5, 6, 7, 8, 9 = *a*	3421519492	*aaaaaaaaaa*	0

0, 1, 2, and 3 will represent allele *A* and the numbers 4, 5, 6, 7, 8, and 9 allele *a* in the next generation (generation two). If the next 10 random number sequence selected is 0010721254, it represents the alleles *AAAAaAAAaa* so that 70% of the alleles are *A*. Table 24.2 represents a repeat of this process for 12 generations. At the twelfth generation, there is a fixation of the *a* allele and, thus, $q = 1$. The frequency of the *A* allele for each generation is plotted in a graph (Figure 24.2) and demonstrates the random fluctuation in frequency until fixation.

Note that in this example the initial frequencies of the two alleles, *A* and *a*, were arbitrarily established as equal (i.e., $p = q = 0.5$). In natural populations, however, not all alleles have equal frequencies. For example, *B* might have a frequency of 0.7 and its allele *b* a frequency of 0.3.

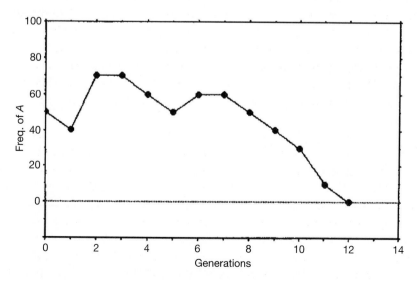

FIGURE 24.2. Random fluctuations in the frequency of the *A* allele until fixation (based on data from Table 24.2).

A.   Procedure

1. Your instructor will provide each pair of students with a random numbers table, which is available in most statistical handbooks. You will use this for calculating changes in allele frequency and completing Table 24.3.

2. For this first exercise, we will assume an initial allele frequency of 0.5 for A and 0.5 for a. Therefore, numbers 0, 1, 2, 3, and 4 represent allele A and numbers 5, 6, 7, 8, and 9 represent allele a.

TABLE 24.3. Simulation Experiment of Genetic Drift Using Random Numbers

Generation	Transforming Random Numbers into Alleles	Random Number	Resulting Alleles	Frequency of A Alleles (percent)
0	—	—	AAAAAaaaaa	50
1				
2				
3				
4				
5				
6				
7				
8				
9				
10				
11				
12				
13				
14				
15				
16				
17				
18				
19				
20				

3. Close your eyes and randomly pick a column and row in the random numbers table. Random numbers tables are usually given in groups of five, so the location where you place your finger represents the first five numbers for generation one. For your next five numbers, move down the column to the next group of numbers.

4. Using 0 through 4 for *A* and 5 through 9 for *a*, determine the number of *A* and *a* alleles generated in the first generation. Enter your data in Table 24.3.

5. Count the number of *A* alleles present in generation one and determine the frequency of the *A* allele. Enter this value in Table 24.3.

6. Now determine the numbers that represent the *A* and *a* alleles in generation two. For example, say your random numbers in the first generation gave three *A* and seven *a* alleles, then 0,1, and 2 will represent the *A* allele and numbers 3, 4, 5, 6, 7, 8, and 9 the *a* allele in the second generation. Enter the values in Table 24.3 for the random numbers you selected.

7. Repeat this process until fixation of one allele occurs.

8. Your instructor has placed a table on the chalkboard in which you are to indicate the allele you fixed and the number of generations required for fixation. Place your data on this table and use the information for answering the questions in Part B below.

9. Use your data and a photocopy of the graph paper on page 305 to prepare a graph comparable to Figure 24.2.

B. Questions Concerning Your Data

1. Which allele, *A* or *a*, was fixed in your experiment? _____

2. How many generations were required to achieve fixation? _____

3. Now, compare your results with those obtained by other pairs of students in the class. Did all

    experiments lead to the fixation of the same allele? _____ Of all the experiments

    done, what fraction or percentage led to fixing the *A* allele? _____ The *a* allele?

    _____

4. Did all experiments require the same number of generations to achieve fixation? _____

    If not, what was the range in the number of generations required for fixation? _____

5. In the following space, discuss the reasons why different experiments varied in terms of which

    allele was fixed and the number of generations required to achieve fixation. _____

    _____

    _____

    _____

    _____

    _____

6. What is the significance for subsequent generations if fixation of an allele has occurred? _____

    _____

    _____

7. Could genetic drift lead to the fixation of a deleterious (harmful) allele in a population?

_____ If so, what might be the consequences of genetic drift for that population?

_____

_____

_____

8. What is the possible significance of genetic drift for the evolution of a species? _____

_____

_____

_____

_____

_____

_____

REFERENCES

BODMER, W.F., and L.L. CAVALLI-SFORZA. 1976. _Genetics, evolution, and man._ San Francisco: W.H. FREEMAN.

CAVALLI-SFORZA, L.L. 1969. Genetic drift in an Italian population. _Scientific American_ 221(2):30–37.

GARDNER, E.J., M.J. SIMMONS, and D.P. SNUSTAD. 1991. _Principles of genetics_, 8th ed. New York: John Wiley & Sons.

GLASS, H.B. 1953. The genetics of the Dunkers. _Scientific American_ 189(2):76–81.

GILLESPIE, J.H. 2004. _Population genetics: a concise guide_, 2nd ed. Baltimore: Johns Hopkins University Press.

HALLIBURTON, R. 2003. _Introduction to population genetics_, Upper Saddle River, NJ: Prentice Hall.

HAMMERSMITH, R.L., and T.R. MERTENS. 1990. Teaching the concept of genetic drift using a simulation. _The American Biology Teacher_ 52(8):497–499.

KLUG, W.S., and M.R. CUMMINGS. 2006. _Concepts of genetics_, 8th ed. Upper Saddle River, NJ: Prentice Hall.

METTLER, L.E., T.G. GREGG, and H.E. SCHAFFER. 1988. _Population genetics and evolution_, 2nd ed. Englewood Cliffs, NJ: Prentice Hall.

ROHLF, F.J., and R.R. SOKAL. 1994. _Statistical tables_, 3rd ed. New York: W.H. Freeman and Co.

RUSSELL, P.J. 2006. _iGenetics: a Mendelian approach_, New York: Benjamin/Cummings Publishing Company.

SOKAL, R.R. 2005. _Biometry_, 4th ed. New York: W.H. Freeman and Co.

ZAR, J.H. 1998. _Biostatistical analysis_, 4th ed. Upper Saddle River, NJ: Prentice Hall.

Note: All references cited in Investigation 23 are also pertinent to this investigation.

INVESTIGATION 25

Applied Human Genetics

Although human genetics is probably the oldest area of applied genetics and *Homo sapiens* is certainly an important organism for genetic study, human genetics developed slowly until the last half-century. In the first decade of the twentieth century, following the rediscovery of Mendel's principles, considerable interest was shown in human genetics. Many traits were found to be more prevalent in certain families than in the general population. Some were initially explained in simple Mendelian terms. "Feeblemindedness," for example, was considered to be a single genetic entity and was interpreted on the basis of a single-gene substitution. Now it is known that only a few conditions that result in mental retardation depend on single gene substitutions and that many types of mental retardation are genetically and environmentally complex. Critical analyses have indicated that most of the early explanations for genetic mechanisms of human traits were oversimplified.

After the first rush of interest in human genetics early in the 1900s, interest waned because human inheritance appeared to be too imprecise and difficult to analyze. It was easier and more productive to apply the experimental method to other animals and to plants. Controlled matings among experimental organisms were preferred over uncertain human studies as the way to establish basic genetic principles. Fortunately, genetics experiments using organisms that were more amenable to the experimental approach led to the discovery of principles that were universally applicable and thereby indirectly furthered human genetics.

Many types of experiments simply cannot be performed on humans, and objective data are elusive even when experiments are possible. Ideally, genetic investigations employ organisms having standardized genotypes grown in a controlled environment. Humans are far from standardized genotypically, and to a large extent they regulate their own environment. Circumstances at home and at school, nutritional status, communicable diseases, and a number of other factors influence the development of a child. Finding anything in human society that resembles the ideal conditions required for objective experimental genetics is very unlikely. Identical twins and other combinations of multiple births provide the only human units that approach a genotype standard. Identical twins and similar multiple births occur infrequently, and usually the members of twin pairs or other multiple birth groups are raised in the same home, subject to essentially the same environmental conditions. When identical twins are separated early in life and experience different environments, one may compare the effects of these environments on similar genotypes.

The study of human genetics has been complicated further by the long period (on the average, about 20–30 years) between generations. The life span of the investigator is no greater than that of the organism being studied. Only recently have methods been developed that allow the gathering of significant statistical data from one or two generations. Furthermore, large families are desirable for genetic studies, and, with rare exceptions, even the largest human families fall short of the size necessary to establish genetic ratios and pursue orthodox statistical analyses.

Other difficulties encountered by human geneticists have been associated with incomplete knowledge of human cytology and the genetic mechanism itself. Since 1956, human cytogenetics has matured into an active and progressive field of research (see Investigations 9 and 10). Of necessity, the emphasis in human genetics has been on phenotypes rather than genotypes. Obviously, a true genetic picture must be obtained before gene behavior can be properly analyzed and evaluated. Hereditary deafness, for example, may result from any one of several abnormalities in the ear, nerve tracts, or brain, with each abnormality controlled by a different combination of genes. By contrast, certain manifestations of a single gene may result in strikingly different phenotypes.

In addition, different genetic mechanisms can produce identical phenotypes in different families. For example, phenotypically similar forms of albinism may be inherited in an autosomal dominant, an autosomal recessive, or an X-linked pattern. These difficulties are being overcome by more precise information concerning gene action.

Many human traits result from the cumulative action of **polygenes** (multiple gene loci), which have yet to be identified or localized individually. Polygenes resemble Mendelian genes in transmission but are cumulative in action (see Investigation 22). Much of the variation known to occur in humans is probably based on numerous polygenic differences for which the separate genes cannot be studied individually.

In spite of these difficulties, significant advances have been made in human genetics, combining more traditional approaches (analysis of phenotypes within family pedigrees) with more recent advances in molecular genetics. Today's investigation looks at these different aspects and the role of genetic counselors in human genetics.

OBJECTIVES OF THE INVESTIGATION

Upon completion of this investigation, you should be able to

1. **list** at least 10 human single-gene traits and indicate the mechanism of inheritance (autosomal dominant, autosomal recessive, X-linked dominant, X-linked recessive) of each,

2. **prepare** and **analyze** a human pedigree illustrating the pattern of inheritance of a single-gene trait,

3. **write** a paragraph outlining the functions of a genetic counselor as these functions relate to detection of heterozygous carriers and to prospective and retrospective genetic counseling,

4. **discuss** the consequences of consanguinity on the risk of bearing a child expressing a rare recessive genetic defect, and

5. **discuss** the electrophoresis of hemoglobin variants and the significance of this procedure for humans.

Materials needed for this investigation:[1]

For 2–8 students:

electrophoresis chamber

applicators

sample well plates

Eppendorf tubes

microdispenser and tubes or micropipette and tips

staining set

cellulose acetate plates

buffer (Super Heme)

hemo AFSA$_2$ control

Ponceau S stain

filter or 3M paper blotters

---

[1]   A Super Z-12 Hemoglobin Kit, which contains all equipment and supplies necessary for this investigation, is available from Helena Laboratories, P.O. Box 752, 1530 Lindbergh Drive, Beaumont, TX 77704-0752.

5% acetic acid

Sharpie marker

Additional supplies if fresh blood is to be used:

hemolysate reagent

blood sample preparation dish (depression slides), small Eppendorf tubes or pieces of Parafilm

toothpicks

70% ethanol

cotton

sterile lancets

I. MECHANISMS FOR GENETIC TRAITS

The classical study of single-gene effects was by conventional pedigree analysis. This procedure is still a valid and useful method for studying many traits. McKusick (OMIM, December 6, 2005) listed 16,419 human traits, many of which were first identified by conventional genetic methods. Of these 16,419 traits, 15,384 are autosomal, 916 X-linked, 56 Y-linked, and 63 are assigned to the mitochondrial chromosome. Many other gene loci have been assigned to specific chromosomes: 8,826 to autosomes, 549 to the X-chromosome, and 44 to the Y. For example, the gene determining wet/dry earwax (see below) is located on chromosome 16, in the centromere region (16 p11.2–q12.1) (See Investigation 10 to refresh your memory on chromosome nomenclature.). The Human Genome Project has resulted in determining a known base-sequence for 10,847 of the 16,419 genetic traits. Not surprisingly, chromosome 1, the longest autosome, has the most assigned gene loci—890, and chromosome 21, one of the shortest, has the fewest loci—127. A useful project for you to do would be updating the statistics in this paragraph by checking OMIM (see References for the Web-site address).

A. Readily Detected Human Traits

Table 25.1 provides a list of readily distinguishable human traits that most students should be able to characterize for themselves. From McKusick's 1998 catalog or OMIM Online or other reliable sources, determine the mechanism of inheritance along with complications and complexities associated with the readily observed (mostly morphological) traits listed in Table 25.1. Use the McKusick number given in the table to locate the trait in the catalog or online (see References). Typically, any trait having a McKusick number beginning with 1 is due to an autosomal dominant, those starting with 2 are autosomal recessive, and those starting with 3, are X-linked.

B. Human Traits Having Biochemical Implications

1. Variation in cerumen (earwax) [McKusick no. 117800].

    Some people have wet, sticky, brownish-colored earwax and others have dry, crumbly, gray earwax.

    What is your phenotype for this trait? _____

    By a date set by your instructor, prepare a pedigree record for your family or a family known to you for this trait and discuss it with your instructor.

    Results of a survey of this trait in a genetics class at Ball State University are summarized in Tables 25.2 and 25.3. One family pedigree chart, that is, a diagrammatic representation of the family, is given in Figure 25.1.

TABLE 25.1. Some Readily Detected Human Genetic Traits

Trait (McKusick Number)	Inheritance Mechanism with Complications and Complexities
Albinism I (203100)	
Baldness (109200)	
Bent little finger (114200) (camptodactyly)	
Brown hair (113750)	
Cleft chin (119000)	
Color blindness (303800 and 303900)	
Dimples (126100)	
Double-jointed thumbs (274200)	
Ear lobes (128900) (free or attached)	
Ear point (124300)	
Eye color I (227240)	
Index finger (136100) (shorter than ring finger)	
Hair whorl (139400)	
Handedness (139900) (left or right)	
Hand clasping (139800)*	
Middigital hair (157200)	
Myopia 2 (160700)	
Polydactyly (174200)	
PTC tasting (171200)	
Red hair 2 (266300)	
Tongue folding or tongue rolling (189300)*	
Widow's peak (194000)	

*See Martin (1975) and Barnes and Mertens (1976) as well as McKusick's Online Mendelian Inheritance in Man, http://www.ncbi.nlm.nih.gov/omim/

In a pedigree chart, the circles represent females and the squares represent males. A straight horizontal line joins the symbols for the mother and father, and a vertical line connects with another horizontal line, the family line, from which the children are represented in order of birth. Roman numerals, if present, designate generations.

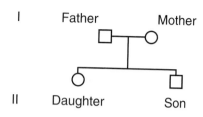

TABLE 25.2. Frequencies of Phenotypes for Cerumen Type among Students at Ball State University

Phenotype	Total	Percentage
Wet, sticky cerumen	79	96.3
Dry, crumbly cerumen	3	3.7
Totals	82	100.0

TABLE 25.3. Familial Data Obtained from Pedigree Studies of Families among Whom Variation Occurred with Respect to Type of Cerumen

Parents	Number of Families	Children		
		Wet-Sticky	Dry-Crumbly	Totals
Wet × wet	28	78	5	83
Wet × dry	16	32	11	43
Dry × dry	1	0	2	2
Totals	45	110	18	128

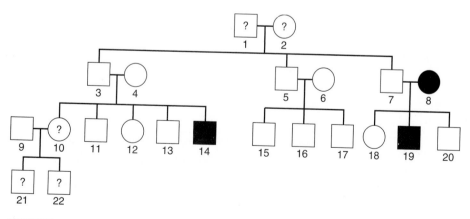

FIGURE 25.1. Pedigree illustrating the typical pattern of inheritance of wet, sticky versus dry, crumbly cerumen. Shaded circles represent females having dry, crumbly earwax, and shaded squares represent males having the same phenotype. Open circles (females) and squares (males) represent individuals having wet, sticky cerumen. Question marks indicate unknown phenotype.

a. On the basis of the information available, postulate the mechanism of inheritance for cerumen type.

_____

_____

Using the data in Table 25.2, calculate the allele frequencies (p and q) and genotypic frequencies ($p^2 + 2pq + q^2$) for the population represented by the Ball State sample (see Investigation 23 for details on the Hardy-Weinberg equilibrium).

b. p and q: _____

c. $p^2 + 2pq + q^2$ _____

2. Detection of S-methyl thioesters excreted after eating asparagus [McKusick no. 108390].

   After eating asparagus, people excrete in their urine the malodorous substances _S_-methyl thioacrylate and _S_-methyl-3-(methylthio)thiopropionate (see White, 1975). The strong odor of these urinary excretion compounds can be detected by some people ("smellers") after only a few stalks of asparagus have been eaten. Other people ("nonsmellers") lack the ability to smell these sulfur-containing compounds. The exact chemical compound in asparagus that is the source of the odorous sulfur-containing compounds excreted appears to be asparagusic acid (Mitchell, 2001). Some researchers also have suggested that there is a genetic basis to excretion.

   If feasible, sufficient quantities of fresh or canned asparagus will be prepared in the laboratory so that a minimum of six spears will be available for each student to eat during the laboratory period. Be alert to the rather unpleasant odor of the sulfur-containing compounds in your urine during the remainder of the day. (Most "smellers" can readily identify the characteristic odor associated with the _S_-methyl thioesters. If you are a "smeller," you may already know it!)

a. Are you able to smell these _S_-methyl thioesters?_____

   If feasible, prepare a pedigree study of the ability to smell _S_-methyl thioesters for your family or a family known to you. Results of a survey in genetics classes at Ball State University are summarized in Tables 25.4 and 25.5. One family pedigree is given in Figure 25.2.

b. On the basis of the information available, postulate the mechanism of inheritance of this

trait. _____

_____

Using the data from Table 25.4, calculate the allele frequencies (p and q) and genotypic frequencies ($p^2 + 2pq + q^2$) for the population represented by the Ball State sample.

c. p and q: _____

d. $p^2 + 2pq + q^2$: _____

TABLE 25.4. Frequencies of Phenotypes for the Ability to Smell _S_-Methyl Thioesters among Students at Ball State University

Phenotype	Total	Percentage
_S_-methyl thioester smellers	18	22.2
_S_-methyl thioester nonsmellers	63	77.8
Totals	81	100.0

TABLE 25.5. Familial Data Obtained from Pedigree Studies of Families among Whom Variation Occurred with Respect to Smelling *S*-Methyl Thioesters

Parents	Number of Families	Children		
		Smeller	Nonsmeller	Totals
Smeller × smeller	8	17	5	22
Smeller × nonsmeller	17	22	19	41
Nonsmeller × nonsmeller	10	0	31	31
Totals	35	39	55	94

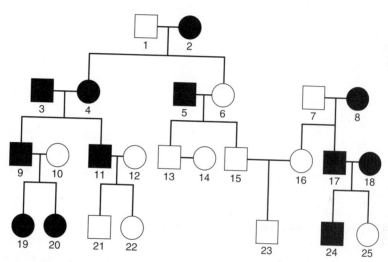

FIGURE 25.2. Pedigree illustrating patterns of inheritance of ability to smell *S*-methyl thioesters excreted in urine after eating asparagus. Shaded circles and squares represent individuals who are "smellers." Data are consistent with the hypothesis that ability to smell *S*-methyl thioesters is due to a dominant allele. Note especially the marriage between individuals 17 and 18, which resulted in the production of individual 25, a nonsmeller.

II. GENETIC COUNSELING

The expression **genetic counseling** suggests the giving of information and support to a counselee and/or his or her family by a group of specially trained individuals. Genetic counseling usually consists of an exchange of information rather than the giving of a directive. Persons seeking genetic counseling are frequently prospective parents who have had an affected child or know of relatives having a hereditary condition. With all of its social, religious, and legal implications, counseling calls for a careful and responsible attitude on the part of the counselor and medical personnel. Initially, genetic counseling was provided by physician-geneticists, but it has evolved over the past several decades to include a number of supporting health groups and specialists. These specialists include individuals specifically trained in graduate master's degree programs for genetic counseling. The first such genetic counseling program was established in 1969 in the United States, and today the American Board of Genetic Counseling has fully accredited 20 institutions in the United States and two Canada.[2] Individuals graduating

[2] The following Web site is for the American Board of Genetic Counseling: http://www.abgc.net/genetics/abgc/abgcmenu.shtml. The Human Genome genetic counseling site is http://www.ornl.gov/sci/techresources/human_genome/medicine/genecounseling.shtml.

from these programs are considered to be certified genetic counselors. According to the National Society of Genetic Counselors (NSGC):

> Genetic counselors work as members of a health care team, providing information and support to families who have members with birth defects or genetic disorders and to families who may be at risk for a variety of inherited conditions. They identify families at risk, investigate the problem present in the family, interpret information about the disorder, analyze inheritance patterns and risks of recurrence, and review available options with the family.[3]

Services rendered by genetic counselors are guided by the NSGC Code of Ethics. This code of ethics states that genetic counselors will help their clients make informed and independent decisions based upon a thorough development of background knowledge and clarification of consequences that possible decisions entail (Benkendorf et al. 1992).

Individuals may seek genetic counseling on reproductive issues, following the occurrence of a genetic condition in a child or adult, or for predisposition or presymptomatic testing for adult-onset conditions. Those seeking genetic counseling provide medical and family history information on the **proband** and his or her relatives. The genetic counselor will construct a pedigree or genetic family history that may assist with making a diagnosis, determining testing strategies, and assessing risks, among other functions. Particular items of special interest to the genetic counselor will vary with the reason for referral. For instance, with reproductive issues, the counselor will assess: **(1) ethnic background** (e.g., high incidences of cystic fibrosis among whites of northern European ancestry, sickle-cell anemia among blacks, Tay-Sachs disease among Ashkenazi Jewish populations, etc.), **(2) parental age** (e.g., incidence of chromosome trisomies increases with maternal age, incidence of certain sporadic dominant single-gene mutations increases with paternal age), **(3) medication use/abuse during pregnancy** (drugs, infections, and trauma can also produce birth defects), and **(4) pregnancy screening** and test results (certain tests like alpha fetoprotein concentrations performed during pregnancy may alter the risk of occurrence of various genetic conditions such as Down syndrome and neural tube defects).

The problem at hand is usually a specific genetic disorder. A reliable prognosis can be made only when the diagnosis has been established as accurately as possible on the basis of clinical and laboratory investigations. Risk assessments for reproductive or personal health issues may be based on established modes of inheritance, empirical risk figures, and/or mathematical calculations incorporating additional information specific to the family. Further testing (amniocentesis for chromosome analysis, carrier testing, etc.) or clinical evaluations of additional family members may be recommended to refine risk calculations. Additionally, referrals to other specialists may be indicated. Throughout the process of gathering and providing genetic information, counselors also acknowledge the psychological issues and responses of clients and assist them via supportive counseling, advocacy, and referrals for community/genetic support organizations.

In summary, the scope of genetic counseling is broad, ranging from preconception, to prenatal, to pediatric, to adult services; and the focus of these services may involve a wide range of genetic and psychosocial issues. In the following sections, we will explore carrier detection, prevention, and the consequences of consanguinity in further detail.

A.   Detection of Carriers

Genetic counseling would be more precise if one could detect heterozygous (carrier) individuals. For example, if a 20-year-old woman with two brothers having X-linked muscular dystrophy asks for genetic advice before starting a family, one could tell her that she has a 50% chance of being a carrier and, in that case, each of her sons would have a 50% chance of being affected. That is a considerable risk. However, if by some laboratory procedure it could be established that she is not a carrier, the counselor could

---

reassure her and send her home rejoicing. Of course, she could be a carrier. In either case, counseling becomes less of a probability estimate than when prediction was based on incomplete knowledge. Methods have been established to detect carriers of a number of traits, for example, Duchenne muscular dystrophy, hemophilia, Tay-Sachs disease, and sickle-cell anemia (see Section III). The vast majority of carrier states still remain beyond detection, although new molecular genetic technology is affording significant advances in this field. (See Investigation 15 on the use of DNA RFLPs in the detection of carriers.) If an individual or couple were identified as carriers of a particular genetic condition, a genetic counselor could provide the family with an accurate risk assessment for having an affected child. Carrier screening is typically available to those with a known or suspected family history of a specific genetic condition for which screening is available and to those who are members of a high-risk subpopulation (see National Society of Genetic Counselors Position Statement on Carrier Identification—www.nsgc.org).

B. Prevention of Serious Genetic Abnormalities

When an individual or couple has the goal of preventing a particular genetic disease, identification of the at-risk genotype is typically the first step. Carriers are often identified through an affected child, affected pregnancy or other affected near-relative, in which case the counseling is retrospective, as opposed to prospective counseling provided in preconception, presymptomatic, and predisposition situations.

1. Prospective genetic counseling.
Although most counseling at present is retrospective, it may provide a family with more options to counsel them before they have an affected child, that is, prospectively. This requires identifying heterozygous individuals by a population screening program and providing appropriate genetic counseling following the results.

The development of techniques suitable for use in mass screening and increased awareness of the significance of this approach for public health policies has led to the initiation of screening programs for carriers of several diseases, and the number is likely to increase as methods improve. Most of these screening programs are voluntary in nature and are limited to populations among whom the frequency and severity of a particular disease are high. There are, for instance, programs for the detection of sickle-cell anemia, glucose-6-phosphate dehydrogenase deficiency (G-6-PD), thalassemia, and Tay-Sachs disease. Furthermore, in 2001, the American College of Obstetricians and Gynecologists (www.acog.org) began recommending that all obstetricians routinely offer voluntary genetic carrier screening for cystic fibrosis to all couples seeking preconception or prenatal care. These programs encompass a commitment to provide genetic counseling for identified heterozygotes. For example, informing individuals identified by a screening program that they are carrying a specific mutant gene that can cause disease in their children is quite a different thing from counseling parents who have already had an affected child. Informed consent prior to testing and genetic counseling following carrier detection are key components to the screening process. The ethical, legal, and social implications of genetic screening are under continuing investigation; in fact, 3–5% of the Human Genome Project budget is devoted to research in this area. Examples of issues that have arisen include recommendations against carrier testing of minors, the occurrence of insurance discrimination and psychological harm, and the discovery of unanticipated information (such as nonpaternity or the revelation of unexpected disease association).

Relatively little is known about the psychological effects of learning that one carries a harmful gene or about the kinds of social pressure that may be brought to bear on individuals so identified. For example, some evidence shows that African-Americans identified as asymptomatic heterozygous carriers of the gene for sickle-cell anemia are unwarrantedly discriminated against because they are carriers. Some heterozygotes who are said to have sickle-cell trait have found their life insurance premiums increased or insurance denied. Others with the sickle-cell trait were denied entry to the U.S. Air Force Academy, apparently on the assumption that if subjected to the stress of flight at high altitudes and low-oxygen tension, their erythrocytes could sickle, thus impairing their effectiveness and endangering their lives and the lives of others. As a result of a successful lawsuit, this latter discriminatory practice has been stopped.

Any mass genetic screening program should be accompanied by a well-designed public education campaign, and the early stages of such programs should include intensive study of the psychological, social, and legal implications of the program. To encourage effective use of genetic counseling services, increased efforts should be made to educate physicians and the general public.

a. Thalassemia, an anemia, is controlled by an autosomal gene (McKusick 604131). When the gene is homozygous (thalassemia major), the disease is severe, with death occurring in childhood. In the heterozygous condition, the gene produces a much milder condition (thalassemia minor). Carriers can be readily detected at a screening center. The disease is prevalent in Cyprus and Sardinia and among other Mediterranean populations.

1. What advice should be given to people of these high-incidence communities?_____

_____

2. What advice should be given to known carriers? _____

_____

3. What educational programs should be initiated for the affected populations? _____

_____

b. A study of three generations in a family of a male colleague of the authors indicated that many of the man's blood relatives had died of cardiovascular accidents and coronary heart disease. Others died of other causes but were affected by high blood pressure. What are the implications of this

information for the man in question? _____

_____

_____

What information and help might a genetic counselor provide? _____

_____

_____

2. Retrospective genetic counseling: The options available to prevent the recurrence of a genetic disorder in a family depend upon many factors, including (1) the availability, accuracy, and cost of detecting the disorder in question, (2) the timing and method of testing, (3) the attitudes and cultural beliefs of those involved, and (4) societal and legal restrictions. The advent of modern reproductive manipulation has also increased the range of choices that carriers of genetic diseases now have for procreation and the prevention of the disease. Consequently, genetic counselors should also be well versed in procedures such as *in vitro* fertilization, gamete selection, and contraception, as well as methods of prenatal diagnosis of fetal conditions. With these considerations and restrictions in mind, one of the following methods or options may be available for consideration. For our discussion we have divided these options into two major categories: pre- and post-conception interventions.

Preconception interventions involve several types of procedures and methods ranging from pre selection of gametes such as sperm selection, to the use of donated eggs, to the more conventional forms of contraception, to far more radical procedures such as ovary transplants and use of a surrogate mother. Each of these procedures either serves to prevent conception or influence conception so as to minimize the risks of a genetic disease.

i. Preselection of gametes. In the case of sperm selection, Y-bearing sperm are lighter and can be separated from X-bearing sperm, although not with 100% efficiency. This has proven useful for increasing the probability that females heterozygous for an X-linked recessive trait will bear female children, thus avoiding the potential of an affected male child. Donor sperm and/or eggs from either an unrelated donor or a pre-screened relative can also be utilized for reducing the risk of passing on a particular genetic condition. Artificial insemination is now a commonly used procedure in our society for infertility and other reproductive options.

ii. Contraception/sterilization. Although no method of contraception is entirely reliable, modern methods are highly effective and are acceptable in many cultures. The most reliable form of contraception would be sterilization and is a choice that many individuals heterozygous for a deleterious gene now elect. Although the term *sterilization* at first seems like a harsh alternative for controlling conception, both tubal ligation in females and vasectomy in males are now commonly practiced forms of birth control.

Postconception interventions can range from utilization of *in vitro* fertilization and embryo selection, to prenatal diagnosis of a genetic condition and pregnancy termination. A large proportion of retrospective genetic counseling is associated with the aspects of prenatal screening and diagnosis of fetal conditions and providing prospective parents with background knowledge about medical conditions and sources for psychological and emotional support. These activities now form the basis from which parents can make informed decisions.

i. Prenatal diagnosis. Over 450 genetic conditions ranging from chromosomal aberrations through single-gene metabolic disorders like Tay-Sachs and certain types of dwarfism can be detected by analysis of fetal cells derived by amniocentesis or chorionic villus sampling. Cost for administering all of these tests for all pregnancies would be prohibitive, but for those individuals with family histories or other risk factors associated with specific genetic disease, the number of screens would be much fewer and the cost would be much more moderate. Baylor College of Medicine has recently initiated a program of prenatal screening for some 50 conditions that cause mental retardation for a total cost of approximately $2000.

When conducting prenatal screening, the NSGC recommends:

> Individuals/couples considering screening should be provided with accurate, balanced information about the condition for which screening is being offered. They should be informed of the specificity, sensitivity, accuracy, risk, benefits and limitations of the screening tests offered and of any follow-up diagnostic tests as well as their reproductive options given a positive diagnostic test result.[4]

ii. Pregnancy termination. This may be an option in cases in which advice is sought after the woman has become pregnant and when amniocentesis or another method of prenatal detection indicates that the fetus possesses a genetic or chromosomal defect.

iii. Preimplantation genetic diagnosis. For those individuals who are known to be heterozygous for a deleterious gene, newly developed procedures associated with *in vitro* fertilization allow assessment and testing of the genetic makeup of early embryos (8-cell stage) through the morula stage. Only embryos known not to possess the disorder are then implanted into the recipient mother, thus allowing heterozygous couples to have a child free of the genetic defect.

---

[4] Position statement from National Society of Genetic Counselors (www.nsgc.org).

a. A 32-year-old woman is the mother of three children, two healthy sons and a daughter 6 years old who has spent much of her life in the hospital with cystic fibrosis, a disease controlled by a single autosomal recessive gene. How should a counselor respond to the woman's request for information and advice concerning the risks for cystic fibrosis in further pregnancies? _____

_____

What are the chances (probability) that her unaffected sons carry the recessive allele for cystic

fibrosis? _____

_____

b. The first child of a normal healthy couple was an albino. What information may be given

concerning their producing albino children in future pregnancies? _____

_____

_____

C. Consequences of Consanguinity

The frequency of recessively inherited diseases among offspring of marriages between relatives is higher than that among children of unrelated couples. If, for example, a person who is known to have inherited a deleterious recessive allele from his great-grandfather were to marry his first cousin, who has a 1/4 chance of carrying the same allele, the risk of producing a homozygous affected child is $1 \times 1/4 \times 1/4 = 1/16$. (A heterozygote by a heterozygote mating has a 1/4 chance of producing a homozygous recessive offspring.) The total risk is thus much higher than that following random mating.

The coefficient of consanguinity (inbreeding) is the probability that any single locus is homozygous by descent from a common ancestor (Table 25.6). On average, each of us carries several harm-

TABLE 25.6. Coefficient of Consanguinity (Inbreeding Coefficient, F)*

For Matings of	F
Father-daughter	$1/4 = 0.25000$
Offspring of identical twins	$1/8 = 0.12500$
Uncle-niece	$1/8 = 0.12500$
Double first cousins	$1/8 = 0.12500$
First cousins	$1/16 = 0.06250$
First cousins once removed	$1/32 = 0.03125$
Second cousins	$1/64 = 0.01563$
Second cousins once removed	$1/128 = 0.00782$
Third cousins	$1/256 = 0.00391$

*First cousins are children of siblings; double first cousins are children whose sets of parents are sets of sibs; first cousins once removed are children from a marriage of a first cousin and a child of his or her first cousin (1 1/2 cousins); second cousins are children from a marriage of two first cousins; second cousins once removed are children from a marriage between a second cousin and a child of another second cousin; third cousins are children of a pair of second cousins. (First cousins share two grandparents in common, second cousins share common great-grandparents, and third cousins share common great-great grandparents.)

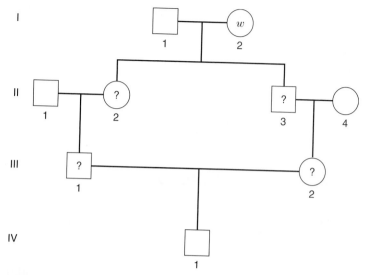

FIGURE 25.3. Diagram of a first cousin marriage to illustrate Wright's method for calculating pathway coefficients of inbreeding from common ancestors. A rare recessive allele (w) is known to be present in I-2 (great-grandmother through both paternal and maternal lines of IV-1, for whom the degree of inbreeding is being calculated). The probability that II-2 is a carrier for w is 1/2. If II-2 carries w, the probability that III-1 also received w is another 1/2, and if III-1 is a carrier, the probability that IV-1 will receive w is another 1/2. Therefore, the probability that IV-1 will receive w from his paternal great-grandmother is $1/2 \times 1/2 \times 1/2 = 1/8$. The probability that IV-1 will receive w through his maternal great-grandmother is 1/2 for II-3. If II-3 is a carrier, the chance is another 1/2 that his daughter (III-2) is also a carrier. If she carries w, IV-1 will have 1/2 chance to receive w from her. The probability for IV-1 to receive w from his maternal great-grandmother is thus $1/2 \times 1/2 \times 1/2 = 1/8$. The probability that IV-1 will receive w from both paternal and maternal lines from I-2 and thus be homozygous (ww) is $1/8 \times 1/8 = 1/64$. The chance is 1/64 (0.01563) that IV-1 will express the trait. Note that it is possible that w could be introduced through II-1 or II-4, but for genes that are rare in the population this probability is very small. The father of the great-grandmother (I-2) expressed the phenotype and was known to have the genotype ww.

ful genes hidden by dominant alleles. Those who have a common ancestry are more likely to carry the same recessive allele than are those from random mating in the general population. Because individuals homozygous for a recessive gene will express the phenotype, and because most mutant phenotypes are deleterious, a risk is associated with marriage among close relatives. The late Sewall Wright devised a method of path coefficients to calculate risks in terms of probability. This method is illustrated in Figure 25.3.

1. A couple of first cousins living in a state where marriage of first cousins is legal has requested information concerning the risks to their prospective children. What information could be given?

2. An adoption agency requested information concerning the likelihood of abnormalities in the unborn child of an incestuous brother-sister mating. What information could be given?_____

_____

_____

III. HEMOGLOBIN ELECTROPHORESIS

Human blood is a complex liquid tissue consisting of cells of different types suspended in a liquid matrix called the **plasma**. One type of blood cell is the **erythrocyte** (red blood cell), which functions in transporting oxygen to the body's cells. The red pigment **hemoglobin** in the erythrocyte is the specific carrier of oxygen. Hemoglobin is a complex iron-porphyrin-protein molecule: Normal adult hemoglobin (HbA) is composed of four polypeptide chains, two identical **alpha** chains each consisting of 141 amino acid residues and two identical **beta** chains each consisting of 146 amino acids. One gene (on chromosome 16) determines the amino acid sequence in the alpha chain, whereas a different gene (on chromosome 11) controls the beta chain. Normal adult hemoglobin, HbA, may be symbolized as $\alpha_2^A \beta_2^A$, which indicates that it consists of two alpha (α) and two beta (β) chains of the normal adult (A) type. A minor form of adult hemoglobin, hemoglobin A_2 (HbA$_2$), consists of two alpha and two delta chains, whereas fetal hemoglobin (HbF) consists of two alpha and two gamma chains. Delta and gamma are different polypeptides that substitute for beta in these hemoglobins.

Gene mutations are known that modify the amino acid sequence of both the alpha and the beta chains of hemoglobin. Certainly the best known of these is sickle hemoglobin (HbS = $\alpha_2^A \beta_2^S$), which is due to an alteration in the DNA that codes for the beta chain. As a consequence of that alteration, the glutamic acid (a charged amino acid) normally found in the sixth position in the beta chain of HbA is replaced by valine (an uncharged amino acid). Individuals homozygous for the sickle-cell gene (Hb^S/Hb^S) (McKusick no. 141900, variant .0243; see also no. 603903) have a serious disease that is likely to reduce their life expectancy; heterozygotes (Hb^A/Hb^S) are essentially symptom free, with a normal life expectancy. The alteration in hemoglobin structure produced by the Hb^S gene reduces the oxygen-carrying capacity of the hemoglobin molecule and causes the erythrocyte to assume a crescent or sickle shape upon giving up oxygen. Such sickled cells are rigid and incapable of passing through tiny capillaries, which then become blocked, thereby depriving body parts of their required blood (oxygen) supply. A painful crisis can result.

A reliable means of identifying heterozygous carriers of the Hb^S allele would be useful in counseling those individuals of their potential for producing offspring having sickle-cell anemia. Because changes in amino acid content can alter the electrical charge on the polypeptide of which the amino acid is a part, such changes can be detected by the process of **electrophoresis**. This, indeed, is the case for sickle hemoglobin and for other mutant forms of the hemoglobin molecule (e.g., HbC, which produces a milder form of anemia called hemoglobin C disease, McKusick no. 141900, variant .0038).

Electrophoresis can be used to separate a mixture of charged molecules in an electrical field. Substances to be separated must bear a net negative or net positive charge, so that they will migrate in a liquid serving as a medium for conducting an electrical current. The mobility of a molecule in electrophoresis will be determined by (1) the voltage applied, (2) the charge on the molecule, and (3) the molecular friction of the molecules involved. Speculate on the following questions and suggest possible answers to them.

1. Would the rate of movement be expected to increase or decrease as the applied voltage is increased? _____

2. Would the rate of movement be expected to increase or decrease as the net charge on the molecule increases? _____

3. Would increased molecular friction (which depends on size and shape of the molecule) be expected to increase or decrease the rate of movement? _____

Cellulose acetate provides a spongy matrix on which hemoglobin samples may be placed for electrophoresis. Hemoglobin samples will be placed on cellulose acetate plates that have been moistened in an appropriate buffer. All student groups will be given a standard consisting of a mixture of different hemoglobins (e.g., HbA, HbS, HbC, HbF) as well as one or two unknowns. Students may be asked to electrophorese their own hemoglobin, in which case additional directions for collecting and preparing the sample will be given. Figure 25.5 provides a standard for identification of hemoglobin types once electrophoresis has been completed.

A.   Electrophoresis Procedure[5]

1. Soak the required number of cellulose acetate plates by slowly (approximately 10 seconds) and uniformly lowering a rack of plates into a container of buffer (Figure 25.4A) or by using a Bufferizer (Helena Laboratories). Soak plates for at least 10 minutes before use. Prior to soaking the plates, you may code them by marking on the glossy, hard (Mylar) side with a Sharpie marker. Place the mark in the upper left corner so you can use it later to distinguish among multiple samples. The left side will later correspond to the side where the first sample is applied.

2. Pour 100 ml buffer into each of the outer compartments of the electrophoresis chamber. Wet two disposable paper wicks in the buffer; then, drape one over each support bridge, making sure it makes contact with the buffer. The chamber is now ready for electrophoresis, but should be kept covered while not in use (Figure 25.4B).

3. The instructor will provide each student with an unknown sample of packed red blood cells for use in electrophoresis or with an unknown hemoglobin sample. (Alternatively students may want to test their own hemoglobin following procedure B listed below.) Add 1 part of packed RBCs to 6 parts of hemolysate reagent in a 1.5-ml Eppendorf or other small test tube. Mix thoroughly by swirling, and let stand at least 5 minutes prior to use. Mix again immediately before use. If you were provided with an unknown hemoglobin sample, it will be ready for proceeding to step 4.

4. Using a microdispenser or micropipette, each student should place 5 μl of hemolysate into one of the wells of the sample well plate (Figure 25.4C). Be sure to place 5 μl of the hemo AFSA$_2$ control in one well. For economy and convenience, do not change the microdispenser bore or micropipette tip with each new sample; merely wash it out in water or buffer. Cover the sample well plate with a glass slide if it is not used within 2 minutes. The easiest way to fill the wells with the microdispenser is to expel the sample as a bead on the tip of the glass tube, and then merely touch this bead to the well. After loading, place a glass slide over the sample well plate to prevent evaporation. Do not start applying the samples to the cellulose acetate plates until all wells are loaded.

5. Prime the applicator by depressing the tips into the sample wells several (three or four) times (Figure 25.4D). The first loading of the samples should be blotted off the applicator by touching the applicator to a strip of blotting paper (Figure 25.4E). This primes the applicator, and the second loading is much more uniform. *Do not reload the applicator at this point but rather proceed to the next step.*

6. Remove the wetted cellulose acetate plate from the buffer with your fingertips and blot firmly between two pieces of blotting paper. Quickly place the strip in the aligning base, with the cellulose

[5] The electrophoresis procedure was developed by Helena Laboratories, P.O. Box 752, 1530 Lindbergh Drive, Beaumont, TX 77704-0752, and is used with permission. All materials listed are available through Helena Laboratories.

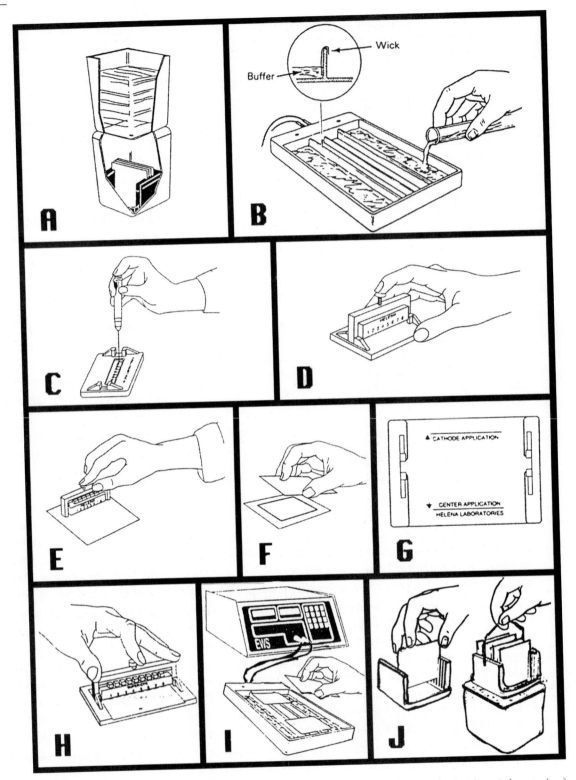

FIGURE 25.4. Steps used in electrophoresis. See text for explanation. (Courtesy of Helena Laboratories.)

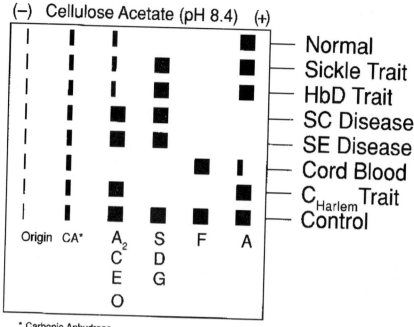

FIGURE 25.5. Distribution of hemoglobins produced by electrophoresis. (1) Normal hemoglobin includes mostly HbA, but also some HbA$_2$; (2) individual with sickle-cell trait has mostly HbA and HbS, but also some HbA$_2$; (3) individual with SC disease has HbS and HbC; (4) cord blood indicates HbA and HbF. For the diagram, the enzyme carbonic anhydrase was used as an internal control found in all hemoglobin types. Certain other hemoglobin variants not discussed in this investigation are also shown. Predict the hemoglobin types you would expect to find in persons homozygous for Hb^S or Hb^C. (Courtesy of Helena Laboratories)

acetate (dull) side up (Figure 25.4F). One end of the plate should be aligned with the back of the aligning base, so that the application spots will be 1 inch away from the cathode (Figure 25.4G). The Sharpie mark should be on the upper left side nearest the cathode. From this point on, sample 1 will always be on the same side as the mark, and sample 8 will be on the opposite side.

7. Slowly depress the applicator tips three or four times into the sample wells to ensure maximum loading of the applicator. Then transfer the applicator to the aligning base. Press the button down and gently hold for 5 seconds (Figure 25.4H).

8. Quickly place the cellulose acetate plate, cellulose acetate side down, in the electrophoresis chamber with the application site nearest the cathode (negative) side (Figure 25.4I). Place a weight (e.g., glass slide) on the plate. Cover the chamber, adjust the voltage to 350 volts, and perform electrophoresis for 25 minutes. *Warning: Electricity is dangerous. Do not touch the electrophoresis chamber or buffer while the power supply is on.* Power must be applied within 5 minutes after a plate has been placed in the chamber. After electrophoresis, turn off the power supply and remove the cellulose acetate plates from the chamber.

9. Place the plate(s) into a staining jar containing Ponceau S (Figure 25.4J), and stain for 5 minutes.

10. Destain in three successive washes of 5% acetic acid for 2 minutes each, gently agitating the plate up and down to facilitate removal of unbound stain. At the end of these washes the background of the plate should be white or almost white. The plates may be dried and stored as a permanent record at this point.

After drying your stained cellulose acetate plate, visually examine the staining pattern on the plate to determine the types of hemoglobin present in your sample by comparing it with Figure 25.5 and with the patterns in the hemo AFSA$_2$ control. *Note:* The quantity of each type of hemoglobin present on the plates can be determined if a densitometer is available. If densitometer readings are to be made, then the plates must be cleared by using one of several procedures (see Helena Laboratories, 1995).

1. List the types of hemoglobin present in your sample._____

_____

_____

2. What types of hemoglobin would you expect in an individual with sickle-cell trait? _____

_____

   Sickle-cell anemia? _____
3. What types of hemoglobin would you expect to find in a blood sample drawn from the umbilical

   cord? _____

_____

B.   Procedure for Fresh Blood

If individual students test their own hemoglobin, they should use the following procedure for the preparation of blood. *Follow all precautions listed in Investigation 23 when handling human blood.*

1. Clean and sterilize fingers with a piece of cotton dipped in 70% ethanol.
2. Prick your finger with a sterile lancet provided by your instructor.
3. Place several drops of whole blood in the sample preparation dish or on a piece of Parafilm and add 3 drops of hemolysate for each drop of whole blood. Mix by gently swirling or by stirring with a toothpick and let stand for at least 1 minute. Mix again immediately before use.
4. Proceed with step 3 in Section A above.

REFERENCES

ALLISON, A.C., and K.G. MCWHIRTER. 1956. Two unifactorial characters for which man is polymorphic. *Nature* 178:748–749.

BAKER, D.L., J.L. SCHUETTE, and W.R UHLMANN, eds. 1998. *A guide to genetic counseling.* New York: WILEY-LISS.

BENKENDORF, J.L., N.P. CALLANAN, S. SCHMERLER, and K.T. FITZGERALD. 1992. An explication of the National Society of Genetic Counselors (NSGC) code of ethics. *Journal of Genetic Counseling* 1(1):31–39.

BANK, A., J.G. MEARS, and F. RAMIREZ. 1980. Disorders of human hemoglobin. *Science* 207:486–493.

BARNES, P., and T.R. MERTENS. 1976. A survey and evaluation of human genetic traits used in classroom laboratory studies. *Journal of Heredity* 67(6):347–352.

BENNETT, R.L. 1999. *The practical guide to the genetic family history.* New York: Wiley-Liss.

CUMMINGS, M.R. 2005. *Human heredity: principles and issues,* 7th ed. Pacific Grove, CA: Brooks/Cole Publishing Co.

HAMES, B.D., and D. RICKWOOD, eds. 1999. *Gel electrophoresis of proteins: a practical approach* (Practical Approach Series). New York: Oxford University Press.

HARPER, P.S. 2004. *Practical genetic counseling,* 6th ed. London: Arnold Publishers.

Helena Laboratories. 1995. *Hemoglobin electrophoresis procedure.* Procedure No. 15. Beaumont, TX.

KHOURY M.J., L.L. MCCABE,-- and E.R.B. MCCABE. 2003. Population screening in the age of genomic medicine. *Genomic Medicine* 348: 50–58.

KOH-ICHIRO YOSHIURA, et al. 2006. A SNP in the ABCC11 gene is the determinant of human earwax type. Published online *Nature Genetics,* January 29, 2006, (doi:10.1038/ng1733).

LEWIS, R. 2007. *Human genetics: concepts and applications*, 7th ed. Dubuque, IA: McGraw-Hill.

LISON, M., S.H. BLONDHEIM, and R.N. MELMED. 1980. A polymorphism of the ability to smell urinary metabolites of asparagus. *British Medical Journal* 281:1676–1678.

MARTIN, N.G. 1975. No evidence for a genetic basis of tongue rolling or hand clasping. *Journal of Heredity* 66:179–180.

MATSUNAGA, E. 1962. The dimorphism in human normal cerumen. *Annals of Human Genetics* 25:273–286.

MCKUSICK, V.A., et al. 1998. *Mendelian inheritance in man: a catalog of human genes and genetic disorders*, 12th ed. Baltimore: Johns Hopkins University Press.

MERTENS, T.R. 1990. Using human pedigrees to teach Mendelian genetics. *The American Biology Teacher* 52(5):288–290.

MERTENS, T.R., J.R. HENDRIX, and K.M. KENKEL. 1986. The genetic associate: a career option in genetic counseling. *Journal of Heredity* 77(3):175–178.

MITCHELL, S.C. 2001. Food idiosyncrasies: beetroot and asparagus. *Drug Metabolism and Disposition* 29(4, part2):539–543.

NICHOLSON, D., and T.R. MERTENS. 1974. Human genetic traits in laboratory instruction. *The American Biology Teacher* 36(8):463–467.

NUSSBAUM, R.L., R.R MCINNES, and H.F. WILLARD. 2004. *Thompson & Thompson genetics in medicine*, 6th ed. Philadelphia: W. B. Saunders.

Online Mendelian Inheritance in Man, OMIM™. McKusick-Nathans Institute for Genetic Medicine, Johns Hopkins University (Baltimore, MD) and National Center for Biotechnology Information, National Library of Medicine (Bethesda, MD), 2000. http://www.ncbi.nih.gov/omim/

VEACH, P. McC., B.S. LEROY, and D.M. BARTELS. 2003. *Facilitating the genetic counseling process: a practice manual.* New York: Springer.

WHITE, R.H. 1975. Occurrence of *S*-methyl thioesters in urines of humans after they have eaten asparagus. *Science* 189(4205):810–811.

WRIGHT, S. 1922. Coefficients of inbreeding and relationship. *American Naturalist* 56:330–338.

YOUNG, I.D. 2000. *Introduction to risk calculation in genetic counseling*, 2nd ed. New York: Oxford University Press.

INVESTIGATION 26

NCBI and Genomic Data Mining

The advent of various molecular techniques such as restriction fragment analysis, cDNA production, cloning, and DNA sequencing led to the acquisition of large quantities of DNA sequence data and the chromosome locations of genes from various organisms. These developments, coupled with the initiation of the Human Genome Project, meant that a systematic reporting mechanism and data storage system needed to be developed and organized so that data could be collected, centrally stored, and easily retrieved for further analysis. Claude Pepper, the late Democratic congressman from Florida, recognized the significance of this problem and sponsored legislation that created the National Center for Biotechnology Information (NCBI) in 1988. NCBI was charged with the mission of developing and coordinating computational biology systems for the acquisition, storage, retrieval, and analysis of molecular biological data. The founders and developers of NCBI designed and implemented a new search and retrieval system called *Entrez*, which forms the basis of many of the NCBI databases. In 1992, NCBI was charged with maintaining and regulating GenBank (a DNA sequence repository) where unique DNA sequences determined by individual research and industrial laboratories can be submitted for storage. In addition to the NCBI databases, there are a number of other national and international databases that are also now fully integrated with NCBI. These include the Cancer Genome Anatomy Project (CGAP), Online Mendelian Inheritance in Man (OMIM), the DNA Database of Japan (DDBJ), and the European Molecular Biology Laboratory (EMBL). The seven major overall goals of NCBI will be introduced to you later, but basically what NCBI does is summarized by the following quote:

> The challenge is in finding new approaches to deal with the volume and complexity of data and in providing researchers with better access to analysis and computing tools to advance understanding of our genetic legacy and its role in health and disease.

The development of the computer systems for the storage and retrieval of this massive amount of molecular data has also led to the development of new fields of inquiry, including bioinformatics, proteomics, pharmacogenomics, and transcriptomics. In this investigation you will be introduced to several major features of the NCBI databases and how they can be accessed and utilized for analysis of gene structure, expression, and cytogenetic location.

OBJECTIVES OF THE INVESTIGATION

Upon completion of this investigation, you should be able to

1. **discuss** the organization and role of NCBI in the acquisition and dissemination of molecular data generated internationally,

2. **access** the PubMed database for citations of previous and current research,

3. **access** NCBI databases on a specific gene's nucleotide sequence and its cytogenetic location on a specific chromosome(s),

4. **analyze** potential alternate mRNA processing of a gene, and

5. **analyze** protein homologies and 3D structure between related proteins.

Materials needed for this investigation

For each student or pair of students:

PC or Mac computer with the following system requirements:

Windows 95, 98, Me, NT, 2000XP

MacOS 10.1 or higher

Downloaded Cn3D 4.1 program for PC, Mac, or Unix

free download: www.ncbi.nlm.nih.gov/Structure, then click on Cn3D to go to the area for the download

Optional application(see Section IV): Gene Workbench; free download: www.ncbi.nlm.nih.gov/projects/gbench

Note. Instructors can have the students complete this investigation in the classroom laboratory if sufficient computers are available for the class and the appropriate software has been downloaded onto the computers. In some cases, it might be more appropriate for the instructor to arrange for this investigation to be conducted in a university's computer laboratory but remember the software will need to be installed by university personnel prior to the laboratory. As a third option, instructors might demonstrate some of the components of this investigation during class time and then have the students work on their own computers at home. Students would then be responsible for the installation of their own software. Finally, the databases at NCBI are continually changing due to additions of new data. This may result in some alterations of the values and alignments that you obtain in this investigation.

I. ORGANIZATION AND ROLE OF NCBI

The designers and staff of NCBI have worked very diligently on designing a database system that is well organized and relatively simple to use for research, given a little background knowledge and a few tips. They have also designed a set of online tutorials for each of the major data analysis and retrieval systems, which will allow you to expand on the things that you will learn in this investigation. In this section, we will introduce you to the NCBI Web site and several of the features that it provides.

A. NCBI Home Page

1. Type www.ncbi.nlm.nih.gov into your computer browser and press go or return. The page that appears is the NCBI home page. At the very top of this page is the name:

NCBI   National Center for Biotechnology Information.

Just below this title is a row with subtitles such as PubMed, All Databases, Blast, OMIM, etc., and just below those terms is a search box selected to All Databases. Place the mouse arrow over the database pull-down menu and press down on the mouse button to see a large list of databases linked through the NCBI site. Once you become familiar with the various databases you can quickly access them by selecting them from this list and hitting the return button.

For now, however, the best way to learn about NCBI is to look at the list of links printed in yellow on the left-hand column of this Web page. At the top of this column is a link entitled "Site Map" followed by "About NCBI", then "GenBank", etc.

2. Center your mouse arrow over the link "About NCBI" and click the mouse button. This is a link that provides a very good overview of NCBI and its various link sites. If you get lost, this is a good

place to return to in order to get reoriented. (Anytime you get lost you can also return to the NCBI home page by clicking the mouse on the NCBI symbol at the top left of the page.)

Figure 26.1 is a photograph of the "About NCBI" Web page showing seven major links starting with "NCBI at a Glance" through "News".

a. Position the mouse over "NCBI at a Glance" and a list of contents appears in the large oval to the right, starting with "Our Mission". For further reading on any of these topics, simply mouse-click on the major topic (NCBI at a Glance) and the menu appears, then select the items of interest by clicking on them. Click on this link and then "Our Mission" to see what NCBI is doing and accomplishing. After reading through "Our Mission", mouse-click on the previous page (backspace) button of your browser to return to the contents list, and then briefly survey the material under "Organizational Structure" followed by "A Story of Discovery." You should return here and read in detail about these and other topics later. These three short essays will provide you with a good overview of the mission and accomplishments of NCBI.

b. Return to the "About NCBI" page by clicking on the previous page button at the upper left of your browser, then place the mouse arrow over "A Science Primer" link and observe the table of contents in the large oval. You should note that the contents range from "Bioinformatics" through "Phylogenetics". A well-written and concise essay explains each of these topics. If you are unfamiliar with them or need a refresher course, you should ultimately return and read through these essays. To read further, mouse-click on "A Science Primer" and then on the spe-

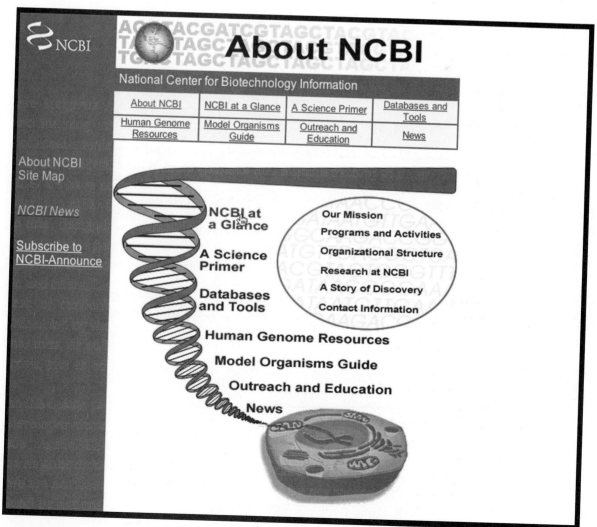

FIGURE 26.1. Photograph of the "About NCBI" web page.

cific topic of interest. When you finish reading, return to "About NCBI" by mouse-clicking on the previous page button.

c. Position the mouse over each of the remaining topics on the "About NCBI" page and observe the contents of each topic for future reference. Note that under "Outreach and Education" the first two items are "Courses and Tutorials" and "Glossaries", which are of special interest to students. Use the "Glossaries" section anytime you find a term that you do not understand, and the "Courses and Tutorials" for developing a more in-depth understanding following completion of this investigation.

 i. Mouse-click on "Outreach & Education" and then select "Glossaries" and click. What appears after clicking on this link are three glossaries (BLAST Glossary, Genetic Glossary, and Genome Glossary) for specialized terms. For example, when we start to compare DNA or protein sequences of two different organisms, we will use a set of programs termed BLAST (Basic Local Alignment Search Tool). Use this glossary for any terms encountered while using BLAST. The "Genome Glossary" includes definitions of many molecular and genomic biology terms.

 ii. Select and mouse click on the "Genome Glossary" link; when the glossary page appears, define the following terms:

Bioinformatics _____

_____

Proteomics _____

_____

Pharmacogenomics _____

Transcriptome _____

_____

At the bottom left side of the glossary terms page is a link entitled "return to home page". Clicking on this link returns you to the "Genome Glossary" page.

 iii. Return to the "Genome Glossary" page by using the above-mentioned link and when the home page appears, take a couple of minutes to observe the additional education links listed on the left-hand-side column. These links can provide important additional Web sites for developing your background knowledge. You are encouraged to return to this site later to explore these additional resources.

 iv. Return to the "About NCBI" by using the previous-page browser button or by using your browser's previous history list. If this fails to return you to the NCBI home page, click the NCBI symbol in the upper left corner and then click on "About NCBI" as you did at the start of this section.

d. Position the mouse arrow over "Databases and Tools" and click to get a list of major databases. The following contents should now appear on the page:

"Literature Databases"
"Entrez Database"
"Nucleotide Database"
"Genomic-Specific Resources"
"Tools for Data Mining"
"Tools for Sequence Analysis"
"Tools for 3-D Structure Display and Similarity Sequences"
"Maps"
"Collaborative Cancer Research"

"FTP Download Sites"

"Resource Statistics"

We will analyze in greater detail several of these databases.

II. LITERATURE AND RESOURCE DATABASES

NCBI has an extensive set of databases in conjunction with the National Library of Medicine (NLM) that can be used for finding citations of research articles, links to full text articles, digital archives of journals, and digitally available books. These resources include PubMed, PubMed Central, and OMIM (Online Mendelian Inheritance in Man). PubMed Central is a site for digitally archived journals maintained by the National Institutes of Health (NIH). Copies of articles published in these journals are available for free downloading. OMIM is a listing of human genetic diseases and genes. For each gene and disease listed in OMIM there is a citation list of important research articles, as well as a concise description of the phenotype. The rest of this section will involve instructions on the use of PubMed for locating research articles on most topics. Most of the citations found in PubMed will have printable abstracts and, for many, full-length articles that can be downloaded for free.

A. Literature Databases

Place the mouse arrow on the "Literature Databases" line that appears on the "Database and Tools"page, and then click the mouse. This opens the "Literature Database" and produces the following list of databases with a short description of each:

PubMed

PubMed Central

Books

Coffee Break

Genes and Disease

Journals

OMIM

Read the short description under each heading and then take a couple of minutes to open each of these databases, PubMed Central through OMIM, to briefly see what is available at each of these sites. Use the previous-page button to return to this page after viewing each database.

B. PubMed Databases

1. Open the PubMed site by mouse-clicking on PubMed. At the top of the page that opens is the PubMed title with its own Web address (www.pubmed.gov). This allows you to go directly to this database without going through NCBI. The dark row immediately beneath the title lists direct links to other NCBI sites, and can be accessed by just clicking on them with the mouse button. Below this line is a search row with the search database set to PubMed, followed by a "for _____" entry space. PubMed can be searched by **author, topic, or journal** following some very simple rules.

2. When searching PubMed by **author**, type the author's last name followed by initials. There should be a space but no punctuation between the last name and initials. For example, cech tr for Thomas R. Cech (a Nobel Prize recipient and highly prolific author). If you are using only the last name of the author without an initial, then place [au] after the last name in order to search only the author category. An example would be cech [au].

a. Type in the "for _____" space **cech [au]** and click go or hit the return button. The Web page that appears has roughly 760 citations (as of February 2006) to various Cech authors in the world. Now try **cech tr** and repeat the process. This time you should find roughly 270 citations that are attributable to T. R. Cech. The difference in the two numbers (490) represents the publications of all of the other Cech authors in the world. Note the small page icon next to each citation. Place the mouse arrow over the page icon and you will notice that a label appears indicating an abstract is available for articles with a two-page icon and free full text article for a multipage icon with green and/or orange on it. A page icon with no lines on it means that no abstract is available.

b. Now look at the second row below where you type the name and you will see a row of terms starting with "Display" with "Summary" in a selection box. Place the mouse arrow on the selection pull-down menu and press on the mouse button. You should see a list of different display options. You can try different displays if you want, but return to the original (default) summary setting for the next step. Also note on the same row there is a "Show" select box, where you can change the number of articles displayed on one Web page, and a "Sort" box, where you can sort the articles by first author, journal, or publication date, etc.

c. Now let's say that you are interested in only the articles that Thomas Cech has published on the telomerase enzyme. What should you do? It turns out that you can limit a search by simply using what are known as Boolean operators—words like AND, OR, NOT. These Boolean operators must be typed as uppercase words. Now try typing **cech tr AND telomerase** into the "search for" space and hit go or return. Note that now only about one fifth of the original Cech articles appear; all on some aspect of a telomerase. Now repeat the above but use **telomerase RNA** instead of just telomerase and only about one tenth of the articles are displayed. Note that a number of these articles are available as free full-text articles. Look at the citation Bryan TM, Goodrich KJ, Cech TR (2003) entitled "Tetrahymena telomerase is active as a monomer" and you will see (if you click on the page icon) that a free full-text article can be downloaded from two sources, one of which is PubMed Central.

d. Next time you write a paper or research report on any topic of genetics, try this database and you will see how easy it is to use compared to most library search systems.

3. To search PubMed for all articles in a **journal**, simply type the full title of the journal or its official abbreviation into the "Search" "For _____" box and press return. For example, type **annual review of genetics** into the search box and press return. You should see a list of all the articles published in each year's volume. If you use **annu rev genet** (no punctuation) you will get the same results. Note that there is an article by Koonin, E.V., 2005 (entitled "Orthologs, paralogs and evolutionary genomics") that you might find very interesting reading.

4. PubMed can also be searched by **"Topic"** following a procedure very similar what we did in section 2c above.

a. In Investigation 25 we discussed a sample of Ball State University students exhibiting variation in cerumen (earwax) type. In this sample 3.7% of the students possessed dry-crumbly cerumen, whereas 96.3% of the students had wet-sticky cerumen. Additionally, filial and pedigree data were presented to suggest that dry-crumbly earwax was inherited as an autosomal recessive trait. (See Tables 25.2 and 25.3 and Figure 25.1 for these data.) Now let's say we are interested in comparing the data that we obtained at Ball State University with other data obtained throughout the world. Obviously, the first thing we need to do is a thorough literature search. PubMed provides an excellent mechanism for searching by topic.

b. To search PubMed by topic, simply type the topic into the "Search" "For _____" space and mouse-click on go or hit return. In this case, type in **earwax** and press return. Write

the number of citations found here_____.

The Web page that appears, as of February 2006, shows some 546 citations for earwax, and these include articles as diverse as earwax removal products to the genetic basis of earwax inheritance. If we repeat this search using cerumen, we find over 600 citations. Clearly, we need to put some limits on the types of earwax citations that we bring forth in our search. We can limit searches in PubMed by simply typing in the "Search" "For_____" space additional qualifying terms that we want to use. Remember, there is no punctuation between words. For example, to limit ourselves to earwax types, let's use **earwax wet dry** as

our search topic. Enter **earwax wet dry** into the "Search" "For _____"

box and press return. Write the number of citations here_____.

   As of February 2006, we found 18 citations. This represents a significant narrowing of the citations. Note that several of the articles deal with the frequency of these phenotypes (and hence genotypes) in different world populations. A reading of several of the abstracts and articles clearly shows that dry-crumbly earwax has a high frequency of occurrence in Mongoloid and Native American populations, but a very low frequency in Caucasian populations; African populations have more intermediate frequencies of the two alleles. Repeat this example using **cerumen wet dry** instead of earwax wet dry and see what you find. Write your results

here_____.

   In our case, based on a search conducted in early February 2006, only 16 citations appeared. This points out something very important. That is, slight differences in terms (earwax versus cerumen) may result in the exclusion of some critical articles that you may want and need. In our example, when we used cerumen as our major term we ended up excluding an article by **Yoshiura, K.I., et al.**, entitled **"A SNP in the ABCC11 gene is the determinant of human earwax type"**. A SNP is a single nucleotide polymorphism. Thus, alleles of the ABCC11 gene are associated with earwax type. The lesson is, always try several different combinations of limiting terms in doing your citation search.

   Boolean operators—words like AND, OR, NOT—can also be used to limit topic searches. Remember to capitalize these words when using them in this fashion.

   Now return to the "About NCBI" page and select "Databases and Tools."

III. MAPS AND MAP VIEWER

In addition to sequencing and annotating the human genome, part of the expanding Human Genome Project has been the sequencing of the genomes of other unique model experimental organisms. Model experimental organisms are those like *Drosophila* that can be easily manipulated genetically, physiologically, and biochemically in order to experimentally answer important biological questions. Since different organisms can be more effectively used for analysis of some types of questions than other organisms, scientists have worked on developing a number of different model organisms. For example, *Drosophila* is very easy to genetically manipulate by standard Mendelian crosses, whereas *Caenorhabditis* (*C. elegans*) is a great organism for investigating developmental events and cell fates. The mouse serves as a good disease-model organism because of its similarity to humans. The "Maps" link under the "Databases and Tools" section of "About NCBI" directs the user to a Web page that serves as a major routing point to the genomic maps of various organisms and several major tools for analysis and comparison of different genomes. In this section, we will use several of these tools and genome maps to analyze gene organization, expression, and homology.

   Open the "Databases and Tools" Web page and click on the "Maps" site. The screen that appears before you will have a number of tools and genome maps listed in blue with a short description of each. Take a minute and scroll down through the list reading the short description of each. In this investigation we will be using "Map Viewer", which is the very first item on the list and the major tool, we will be using. About halfway down the list are "Model Maker" and "Human-Mouse Homology Maps", which we will introduce some aspects of.

A. Map Viewer

Mouse-click on the "Map Viewer" link and a page appears that has a number of organisms divided into categories ranging from Vertebrates to Fungi. Take a minute to scan through the different organisms that are listed here and pay special attention to the vertebrates. Also note that at the bottom of the page are bacteria, organelles, and virus categories.

Next to the scientific name of each of the organisms is the yellow word "BLAST". "BLAST" stands for "Basic Local Alignment Search Tool" and is a program developed for looking for similarities (alignments) between sequences (either nucleotides or amino acids) in the NCBI databases and a sequence that you are interested in or are submitting (we call this the *query*).

1. Using Map Viewer to find a gene in humans.

   At the end of Section II, we found an article published in *Nature Genetics* that identified a SNP (single nucleotide polymorphism) of the ABCC11 gene, the gene responsible for earwax type (dry-crumbly/wet-sticky). Let's say that we are interested in learning where this gene is located, what its genomic sequence is, what its transcript(s) look like, and how its protein is shaped and organized into functional domains. The "BLAST" program and Map Viewer will allow us to answer many of these questions.

   a. Click on the "BLAST" symbol next to *Homo sapiens* and a Web page entitled "Blast Human Sequences" appears. Below the title is a line for "Databases" with its search category set to "genome (all assemblies)". Place your mouse arrow on the select pull-down menu, and press down on the mouse button to reveal the various databases that can be blasted; the last one on the list being SNPs. You may want to return to this SNP database later and look for the differences between the two earwax alleles. Below the Database line is a Program line with its select category set to "megaBlast", which is used to compare highly related nucleotide sequences. Press the mouse button on the pull-down menu and you can see an additional five BLAST programs with an explanation of what each is used for. Note that BLASTn compares nucleotide sequence to nucleotide sequence, and BLASTp compares protein to protein, but BLASTx compares a nucleotide sequence against protein, whereas a tBLASTn compares a protein to a nucleotide sequence. Hence all kinds of comparisons can be made using Blast. The rectangular box is the area where a query entry is placed by typing it in or through a copy and paste routine. The query entry can be an accession number, gene identification number (gi), or sequence. The "Expect" category is a statistical function gauging the stringency of the alignment. The lower the E value, the more stringent the alignment.

   b. Since we are interested in the ABCC11 gene, we can get to its chromosome location by simply entering the gene name (ABCC11) next to the Map Viewer search box at the top of the Blast Human Sequence screen and pressing go or return, instead of using BLAST. Enter this name now and press go or return.

   The Web page that appears is an "Entrez Genome" page showing the 22 autosomes, two sex chromosomes, and the mitochondrion chromosome (MT). When you look at chromosome 16, you may see a red 2 under the chromosome number and two red lines close to the centromere in the q arm. This means that there are two possible locations for this gene, which in this case is actually due to a delay in readjusting map positions. We will see in a few minutes that the cytogenetic and DNA sequence maps for this gene are not in alignment yet. (By the time you look at this, the position may be realigned and only one position may be observed.) This is an important aspect of these maps; that is, they are continually being refined as new data are submitted.

   Had we been "Blasting" the genome with a sequence that was related to a multiple gene family, then several locations on the same or different chromosomes might be expected. When this occurs, different locations may be marked by different colors. Red indicates strong similarity, purple moderate similarity, and green even weaker similarity. Later we see how this can be used to look at homologies between different organisms and their locations. At the bottom of this page are two references for the ABCC11 gene; the NCBI reference sequence and the Celera Corporation sequence.

c. Mouse-click on the "reference" ABCC11 gene symbol and a Map Viewer page of the chromosome region where the ABCC11 is located appears. Figure 26.2 is a photograph of this page showing the ABCC11 gene as well other genes located within a specific area of chromosome 16. The region displayed is given in the upper left (45,850K to 48,870K). The chromosome location for ABCC11 is given as 16q12.1 and is shown in red on the chromosome ideogram on the left side of the page. The position of the gene is also given as the red line on three types of DNA sequence maps located between the ideogram and the gene name. The "Genes_Cyto" is the cytogenetic map. The "HsUniG" is what is called the human unigene clusters map, which is a group of mRNA (cDNA) clones and ESTs (expression sequence tags—short cDNA clones) for a region, and the "Genes_seq" is the gene sequence map. The gene itself is indicated in blue, with the thicker blue lines indicating exons and the thinner lines introns. The black arrows (under O) show the direction of transcription of a particular gene; an arrow pointing down indicates the gene is on the plus strand, and pointing up on the minus strand. Strand position is also indicated by placing the gene (blue) on the right of the map line (thin black line) for the plus strand and on the left for the minus strand. Note that the ABCC11 gene is on the minus strand, whereas the LONP gene is on the plus strand.

Click on the ABCC11 gene symbol, and a Web page opens that gives the gene name, gene ID number, and a summary about the gene and its protein. Note that there are four important features about the protein(s) of this gene: (1) This is an ATP-binding cassette protein that is also a membrane transporter. This means that the protein should have several different domains associated with different functions. (2) A SNP causes the difference between earwax types. (3) This gene and another gene, ABCC12, are derived from a relatively recent (in evolutionary terms) duplication. [When you return to the previous page, you will see that ABCC12 is on the same DNA (minus) strand and adjacent to ABCC11]. (4) There is an alternate processing

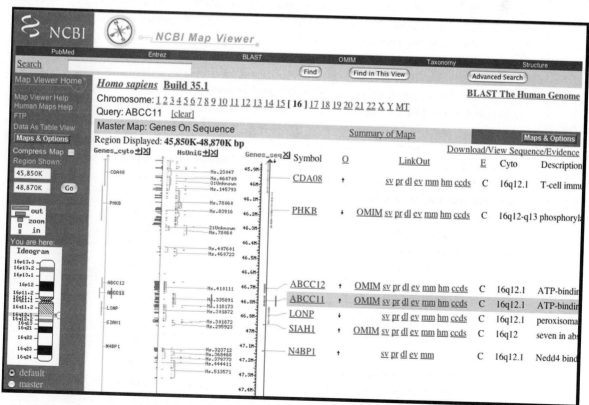

FIGURE 26.2.   Photograph of the NCBI "Map Viewer" web page showing the region of chromosome 16 that possess the ABCC11 Gene Locus.

of the gene to give several variant mRNAs and hence proteins. Additional information on these genes can be obtained by clicking on the OMIM link to the right of the ABCC11 gene symbol. This link will take you to the "Online Mendelian Inheritance in Man" site.

Take a minute and investigate what you can learn about ABCC12 and the SIAH1 genes and briefly record a few pertinent details below. (Don't forget to check OMIM.) _____

_____

_____

_____

_____

When finished, return to the "Map Viewer" page by using the previous-page button.

d. It is possible to look at additional types of maps in Map Viewer and to zoom in and look more carefully at the genes in question. To change the maps on Map Viewer select the yellow "Maps and Options" button on the left-hand side of the page. Click on this button and a screen appears that will allow the addition or removal of various maps. Scroll down the left-hand list and select "Hs_EST" (human EST map). Then mouse-click on the "ADD" button and the "Hs_EST" map appears on the right-hand list. Now select the "Hs_UniGene" on the right-hand list and click on remove; the "Hs_UniGene" is removed from the right-hand list. Now click the "Apply" button at the bottom of the page. The "Hs_EST" map appears and the "Hs_UniGene" disappears from Map Viewer. The "Hs_EST" map shows the alignment of several different clones for these genes and, in essence, represents the exons and introns of some of the variant mRNAs that are produced by these genes.

There are two ways to zoom in (or out) on the genes in question. The first uses a "Zoom out/in" box on the left-hand side of the "Map Viewer" page. Place the mouse arrow over the blue bars and a proportional value of the chromosome appears. Place the arrow on the central bar and a sign "show 1/100 of the chromosome" appears.

Click on the 1/100 bar and observe that the region displayed is now around 46,350K to 47,240K and, though larger, you can still observe several genes.

Repeat this process with the 1/1000 bar and observe the results. Notice at this point the region displayed is about 100K bases and the ABCC11 gene almost fills the screen. In this view, the differences in the exon patterns between variants start to become obvious in the "Hs_EST" map.

The second way to zoom in/out is to place the arrow close to the map line; a hand icon appears. Press down on the mouse button and a list appears with several options starting with recenter. Select the option wanted, then click the mouse.

e. Located to the right of the OMIM link are several other links symbolized by two or three letters. The following is a list of those links with an explanation of what they are for.

sv   sequence viewer giving the sequence on plus or minus strand and transcription areas

pr   information on the protein, including accession number, protein number, gi

dl   DNA sequence download

ev   evidence viewer

mm   model maker for predicting potential mRNA transcripts

hm   homoloGene for looking at gene homologies between different species

We will look at several of these in the following sections.

B.   Model Maker

Model Maker is a program for viewing and analyzing variant forms (isoforms) of the mRNAs that can potentially be formed from different combinations of reputed exons. One of the things that Model Maker does is to allow you to construct your own model of a mRNA by adding or removing exons and then testing to see if there are appropriate ORFs (open reading frames). If you construct a feasible mRNA, then it is possible to predict the protein that would be produced and then test whether any such protein is actually known to exist.

1. **Using Model Maker** You should have the "Map Viewer" Web page on your screen and the zoom set to 1/1000 of the chromosome. If you are not on this view, return to this page by using the previous-page button. Note that at this zoom level you can see the reputed exons in the gene sequence next to the "Hs_EST" map line. The exons are the thicker blue lines. Let's say that we would like to know more about these exons and possible alternate processing to form variant mRNA classes. How might we predict possible alternate forms for the ABCC11 gene, and do we see all of the possible forms in the known clones observed in our current "Map Viewer" page? We can answer some of these questions by using Model Maker.

a. Place the mouse arrow over the **"mm"** symbol to the right of the ABCC11 gene symbol and open the program by clicking the mouse. The "Model Maker" Web page should appear with the organism, chromosome number, contig number, and locus ABCC11 toward the top of the page. Below that should be a series of mRNA (cDNA) clones in blue (approximately 13) that have been isolated and characterized. These are the same mRNAs that you observed in Map Viewer. Below these clones is a graphic view of the putative exons (numbered and in green) for this ABCC11 gene.

   We can construct potential mRNAs by selecting different arrangements of the exons. This can be done by placing the mouse arrow over the number of the exon desired and then clicking. That exon will now appear in the rectangular box below the "Your Model" heading. For example, if you click on exons 4,5,6,7,8,9,10,11,12 you see the mRNA model being constructed in red. If you now add exon 14, you see the red line fork, indicating the presence of a termination codon, which would end the putative polypeptide. This polypeptide would, however, have 553 amino acid residues in it. If you try adding exon 1 or 2, it doesn't add any additional amino acids because it doesn't have a start codon. The DNA codon sequence for this mRNA is below the model, and below that, the open reading frame (ORF) shows 553 amino acids in the model protein. (Frame 1, frame 2 and frame 3 refer to the three potential reading frames.) The amino acid sequence is given in the box below the ORF value. (Each of the different amino acids is indicated by a single capital letter. In the appendix at the end of this investigation is a chart listing the amino acids and their one letter symbol.) The particular model that we constructed corresponds to the third clone in the chart above the exon list. Had we added exon 13 instead of 14, the protein would not have been terminated, and we could have constructed a longer model.

   To the right of the sequence box is a link for an "ORF Finder". If you click on this link it will take you to a page where you can look for alternate reading frames within the protein. Remove exon 1 before going to this link. Mouse-click on the "ORF Finder" and open this page. Click on the first green bar; it turns purple, the first initiation codon is highlighted in pale green, and the termination codon at the end is purple. This is the protein we predicted above. The other green bars represent other possible reading frames, but they all produce fairly short amino acid sequences. Click the mouse arrow on the second bar; it becomes purple and there is a new termination codon only 42 amino acids later. This putative protein would not be very likely as a functional protein. Also note that you can try alternate start codons by clicking the arrow on the appropriate button and repeating the process above.

b. Return to the previous ("Model Maker") page and see now many different models you can construct that have sufficient length to be a possible functional protein. Use the space below to record the exon numbers for the various potential models that you construct.

_____

_____

_____

_____

_____

_____

_____

_____

C.   Homologous Sequences: Paralogs and Orthologs

The major paradigm of biology is that organisms evolve and undergo speciation by the alteration and accumulation of differences in genes. When students think about the evolution of genes, they typically think about the accumulation of mutations in a gene over time. Consequently, different species that arose from a common ancestor should have genes that, while different, show a lot of similarity (homology) with one another. How similar these homologies are should be a function of several things, including how long ago the genes diverged and how much selective pressure there is to conserve the protein structure. Homologous sequences in different species are called **orthologous** sequences when they share a common ancestral gene. While many orthologous sequences in different species have similar functions, in some cases the function may be quite different.

A second way in which we see genes evolving is through gene duplication processes, where a gene becomes duplicated within a species. Once a gene becomes duplicated, then each of the genes has the potential to diverge from one another in their structure and function, thus forming what we call gene families. There are many examples of gene families, hemoglobins and myoglobins being the classic cases cited. Gene duplications in the same species result in homologous sequences we term **paralogous** sequences. In the previous section we discussed the earwax locus and some aspects of its alternate transcripts, but it also turns out to be a good example of a duplicated gene, i.e., ABCC11 and ABCC12. We will look at this duplication a little later.

1. Return to the "Map Viewer" Web page showing the ABCC11 gene locus on the "GenesSeq", "Hs_EST", and "GenesCyto" maps. Mouse-click on the "**hm**" symbol to the right of the ABCC11 locus name. What appears is a Web page entitled **"HomoloGene" (Discover Homologs)**, a site that gives known homologous sequences in other species. If you click on "Discover Homologs", you will be taken to an information page that briefly describes HomoloGene and its automated system and the number of genes from various organisms that have been placed into a homology group. Also notice that at the bottom of the page are several additional links. Of particular interest is the link to **COG**, or Clusters of Orthologous Groups. This Web site gives information on groups of protein paralogs that represent ancient conserved protein domains. Return to the "HomoloGene" home page (the previous page).

2. Below the title for the "HomoloGene" page is a summary of the results generated for the ABCC11 gene that we selected from the "Map Viewer" page. Notice the following HomoloGene listing:

> HomoloGene:69511. Gene conserved in Eutheria
> | | |
> |---|---|
> | H.sapiens | ABCC11 |
> | P.troglodytes | LOC473275 |
> | C.familiaris | LOC478138 |

This indicates that the human ABCC11 gene has orthologs in chimps (Pan) and dogs (Canis). In order to investigate the differences in more detail, mouse-click on the HomolGene number(69511) and a Web page duplicated in Figure 26.3 appears. On the left-hand side of the page, three putative homologs are listed, with the human ABCC11 gene first. Immediately to the right, the protein domains are shown with their sequence order and number of amino acids. The thicker red line indicates the position of the ABC ATPase cassette domain and the thicker gray line, the ABC transmembrane domain. Note that the human and dog ABCC11 have a very similar number of amino acids (1382 vs. 1384) and order, whereas the chimp protein is larger (1545 amino acids) and ordered differently. Just below these protein diagrams is a section entitled "Conserved Domains" that links to the CDD site, with the two domains that we are interested in listed. The select button next to the CDD number acts as a link to the conserved domain site for that protein. We will return to these links later.

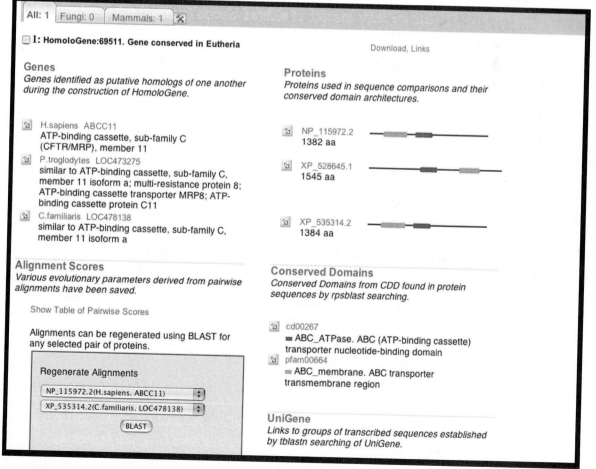

FIGURE 26.3. Photograph of the HomoloGene web page showing homologous chimp and dog gene for the ABCC11 gene locus. Also present on this page are the conserved domains for the ABCC11 gene.

3. On the lower left-hand side of the "HomoloGene" page is an area entitled "Alignment Scores" and below that a rectangular box called "Regenerate Alignments", which is a selection box for performing a "Blast" of selected sequences (currently selected for human and chimp ABC proteins). Click on the "Blast" button to generate an alignment between the human and chimp ABCC11 proteins. The Web page that appears has a graphic alignment of the human protein (top blue line) with the chimp protein (bottom blue line). Below this graphic alignment is the actual amino acid sequence of each. Since it's difficult to see all of the misalignments in these sequences, go to the "View option" above and select "Mismatch-highlighting" and press the align button again. What now appears is the human amino acid sequence in black and the chimp sequence black dots when there is a perfect alignment and **red letters** where there is a misalignment. Note that there are only 19 single amino acid substitutions throughout the protein but three strings (all red) of amino acids in chimp that are not found in humans (black dashes) and one human sequence not found in chimps (red dashes). The thinner red lines on the graphic display above represent these noncomplementary areas. It might be interesting to see if these additional areas are exons that are incorporated into the chimp transcript but not in the human transcript, or if they represent additional nucleotide sequences incorporated into the chimp DNA that are not found in humans.

4. Repeat the steps in 3 above to compare the human and dog protein alignment. Record a summary of what you discover here. _____

_____

_____

_____

_____

_____

In thinking about the human-dog comparisons, you should ask yourself if these comparisons influence how we should envision the human-chimp evolutionary relationships regarding the ABCC11 gene.

D.   Conserved Domains and 3D Structure of Proteins

Conserved domains are areas of proteins in which the amino acid sequence has been maintained through evolutionary history. Many times, but not always, the function of a conserved protein domain is also conserved or at least similar in function. Consequently, when analyzing a new protein, if we find a conserved protein domain, and that domain is typically associated with a specific function, then it implies something important about the overall functioning of that new protein. Additionally, since conserved domains are found in many different species, many of the proteins have been carefully analyzed by X-ray crystallography and their three-dimensional (3D) structure determined. This 3D structural analysis allows predictions about mechanisms of action. In this section we will analyze the conserved domains associated with the ABCC11 gene and relate those domains to its function.

1. If you have not already returned to the "HomoloGene" home page, do so now. Type **ABCC11** into the "search for _____" box at the top of the Web page and press go. What appears is a page showing three groups of related conserved proteins. The first group, which we have already looked at, is most closely related to the ABCC11 gene, whereas the second is related to the ABCC12 gene, and the third to the ABCC10 gene. Click on the HomoloGene number for the first group (ABCC11). This take us to the HomoloGene:69511 page that we used in

Section C.3 above. On the right-hand side of the page is a section entitled "Conserved Domains" and below that are listed two conserved proteins. This means that the ABCC11 gene produces a protein that has two distinct conserved protein domains: (1) an ABC ATP-binding cassette, and (2) an ABC transporter transmembrane region.

2. Click on the number for the human protein (pfam00664) that is found below "Conserved Domains" on the right-hand side of the page. This takes us to the Web page for "Conserved Domains" and the transmembrane protein that we saw in humans, chimps, and dogs. Immediately under the title is a brief description of the domain. You should notice that this protein region consists of a single domain family. On the upper left-hand side of this page are some links to other sites of interest. If you click on "architectures", you will be taken to a site (CDART) where you can see how this protein domain is used in conjunction with other domains for a variety of different functions and in different organisms. Look at the first page of this site and observe the different combinations in which this ABC membrane domain is used. (You may also want to read about CDART by clicking on the "About CDART" link.) When you finish observing this site, return to the "Conserved Domains" page by using the previous-page button.

3. It is also possible to observe a 3D representation of the protein domain. Click on the "Show Structure" button in the lower left-hand side of the "Conserved Domains" page and a download of the 3D structure will occur. After the download finishes, drag and drop (Mac computers) the downloaded image on the Cn3D program icon installed earlier (see Materials). (Windows PCs will automatically open the image.) The download of the 3D structure may take a couple of minutes depending on your Internet connection and how busy the NCBI server is. If you have not already downloaded the Cn3D program, you can download the program now by clicking on the (**download Cn3D**) link.

When the 3D program opens, a diagram of the conserved domain appears as well as two additional (smaller) screens. These screens are entitled "CDD Descriptive Items", which we will not use now, and "ABC_membrane_Sequence/Alignment Viewer", in which the sequence from several different organisms (gi ###) are aligned.

a. Bring the 3D diagram to the front by clicking the mouse on the screen and then look at the Cn3D pull-down menu at the top of your computer screen. Pull down the "View" menu and zoom out a couple of times. Then, using the "View" menu, go down to the "animation" side select screen and select "spin". The model now rotates 360 degrees. (You can zoom in for closer view or zoom out for a more distant view.) Now stop the spinning by using the "View select" menu. You can also change the position of the model by placing the mouse arrow on the structure, depressing the mouse key, and dragging the image in any direction. Do this now.

b. Bring the "Sequence/Alignment" screen to the front and highlight a short row of amino acids in the first row by dragging the mouse arrow over them while holding down the mouse button. (The conserved amino acids are capitalized and in blue or red before highlighting). You should notice that some areas of the 3D structure now become yellow, indicating the position of each of the amino acids highlighted. Even individual amino acids can be visualized in this way. This can be a very useful feature, particularly if you were observing single amino acid substitutions in different alleles. We will also see later how this idea can be used to highlight specific regions of function within the protein.

c. In order to observe where the conserved domain is relative to the entire protein, go to the "Show/Hide" menu and select the "Show Everything" selection. The view that appears shows the whole protein with the conserved domain in color and the nontransmembrane domain sequences in gray. Now reactivate the "Spin" option under "View" and observe what the complete protein looks like and where the conserved regions are relative to the whole.

d. There are different styles for the 3D structure, arranged from "wires" through "space-filled" models. Using the "rendering shortcuts" under the "Style" menu can change these different styles. Take a minute and play around with this and the "coloring shortcuts", where different regions of the protein can be colored.

4. Return to the "Conserved Domains" section of the HomoloGene Web page and repeat steps 2 and 3 above for the ABC_ATPase domain (protein cd00267). This represents the ABC ATP-binding cassette domain. Select this protein by simply clicking on the protein number. You should now see a "Conserved Domain" Web page dedicated to the ABC_ATPase domain. This appears very similar to the transmembrane page except that you now see a section entitled "This domain model appears to be related to other CDs (conserved domains)". Below this, your domain of interest is in red while the other related domains are in green. Read the description of the ABC_ATPase domain and then click the "Show Structure" button. What downloads will be the model of this domain with all of these related domains aligned. This download may again take a couple of minutes. When the download is completed, drag the 3D icon on top of the Cn3D program icon to open. Immediately zoom out several times on the 3D view and then select "Show Everything" under the "Show/Hide" menu. Bring to the front the screen "CDD Descriptive Items" and click on the "Show Annotations Panel". This step produces an additional smaller screen entitled "CDD Annotations", in which the conserved domain is subdivided into different functional components. When you bring this screen to the front, the "ATP binding site" should be selected. Press the "Highlight" button with the mouse and observe the results. Notice that this region is now highlighted in the conserved domain on the 3D model. Repeat this process for these other regions. (Note that you must stop the spinning in order to change the highlighting.)

5. If time permits, go back and repeat Section IIIC and Section IIID using the ABCC12 gene, and the HER2 gene that is shown on the front cover as amplified genes in breast cancer cells. The lines below are provided so that you can record notes and observations on what you found out about

   the ABCC12 and HER 2 gene homologies. _____

   _____

   _____

   _____

   _____

   _____

   _____

   _____

   _____

IV. GENOME WORKBENCH

NCBI has recently developed an integrated computing platform entitled "Genome Workbench" that allows an easy viewing of and coordinated access to all the databases and resources, as well as provides a centralized area for importing and exporting data and analyses of that data. This program runs on your computer and allows projects to be saved for later access and further analysis. A detailed discussion of this application is beyond this investigation, but having now been introduced to the NCBI databases, students should have a sufficient background to download this free program and begin learning about this exceptional tool offered by NCBI. Additionally, NCBI has developed a set of tutorial links as a quick startup and interaction guide for the novice and a set of four detailed tutorials—entitled "Basic Operations," "Non-Public Data," "Working with Multiple Views," and "The Gene Story"—that provides a progressively more complex set of skills in the use of Genome

Workbench. This application can be used with both Windows 2000/XP and above, MacOS 10.3, MacOS 10.4, and Linux/Unix/Source systems. Free downloads of the application are available at **www.ncbi.nlm.nih.gov/projects/gbench**.

APPENDIX

Abbreviations of the 20 Common Amino Acids

One Letter		Three Letters
APOLAR (HYDROPHOBIC)		
Alanine	A	Ala
Valine	V	Val
Leucine	L	Leu
Isoleucine	I	Ile
Proline	P	Pro
Methionine	M	Met
Phenylalanine	F	Phe
Tryptophane	W	Trp
POLAR (UNCHARGED)		
Glycine	G	Gly
Serine	S	Ser
Threonine	T	Thr
Cysteine	C	Cys
Tyrosine	Y	Tyr
Asparagine	N	Asn
Glutamine	Q	Gln
NEGATIVELY CHARGED		
Aspartic acid	D	Asp
Glutamic acid	E	Glu
POSITIVELY CHARGED		
Histidine	H	His
Lysine	K	Lys
Arginine	R	Arg

REFERENCES

ALTSCHUL, S.F., et al. 1990. Basic local alignment search tool. *Journal of Molecular Biology* 215:403–410.

ASHBURNER, M. 2006. *Won for all: How the Drosophila genome was sequenced.* Plainview, NY: Cold Spring Harbor Laboratory Press.

BACEVANIS, A.D., and B.F.F. OUELLETTE, eds. 2004. *Bioinformatics: A practical guide to the analysis of genes and proteins,* 3rd ed. New York: Wiley-Interscience.

CLAVERIE, J.M., and C. NOTREDAME. 2003. *Bioinformatics for dummies.* New York: Wiley Publishing, Inc.

GISH, W., and D.J. STATES. 1993. Identification of protein coding regions by database similarity search. *Nature Genetics* 3:266–272.

MARCHLER-BAUER, A., et al. 2005. CDD: A conserved domain database for protein classification. *Nucleic Acids Research* 33:192–196.

MCGINNIS, S., and T. L. MADDEN. 2004. BLAST: At the core of a powerful and diverse set of sequence analysis tools. *Nucleic Acids Research* 32: 20–25.

MOUNT, D.W. 2004 *Bioinformatics: Sequence and genome analysis,* 2nd ed. Plainview, NY: Cold Spring Harbor Laboratory Press.

PRIMROSE, S.B., and R.M. TWYMAN 2006. *Principles of gene manipulation and genomics,* 7th ed. Williston, VT: Blackwell Publishing.

SCHAFFER, A.A., et al. 2001. Improving the accuracy of PSI-BLAST protein database searches with composition-based statistics. *Nucleic Acids Research* 29:2994–3005.

TISDALL, J. 2001. *Beginning perl for bioinformatics.* Cambridge, MA: O'Reilly Media, Inc.

ZHANG, J., and T.L. MADDEN. 1997. PowerBLAST: A new network BLAST application for interactive or automated sequence analysis and annotation. *Genome Research* 7: 649–656.

SUPPLEMENTAL LABORATORY TOPICS

Instructors wishing to provide individual students with topics for research or special projects may find Supplements 1–3 useful for this purpose.

SUPPLEMENT 1

Plant Speciation by Interspecific Hybridization and Allotetraploidy

Examples of interspecific and even intergeneric hybridization in plants are well documented in the literature. The hybrids resulting from such matings are usually sterile, because the chromosomes of the two species (or genera) are nonhomologous and do not pair properly in meiosis, resulting in nonfunctional microspores (sperm) and megaspores (eggs). Occasionally, however, a rare gamete is produced that has a complete set of chromosomes of each parent. If two such gametes unite, a zygote may be produced that develops into a fertile tetraploid individual, as illustrated in the following sequence: Let $A =$ a haploid set of chromosomes of the one diploid species and $B =$ the haploid set of the second species. The mating $AA \times BB$ will produce the sterile F_1 hybrid AB. If this hybrid produces a few AB gametes, then an $AABB$ hybrid (an allotetraploid) can be produced through union of those gametes. This $AABB$ individual may be fully fertile because A chromosomes synapse with A chromosomes and B with B chromosomes.

Such allotetraploids may occur naturally or may be the product of human intervention. The Russian cytologist G. D. Karpechenko crossed the radish (genus *Raphanus*) with the cabbage (genus *Brassica*), produced the near-sterile F_1 hybrid, and from it obtained a few fertile allotetraploid hybrids (genus *Raphanobrassica*). *Raphanus* ($2n = 18$) and *Brassica* ($2n = 18$) are different genera in the mustard family (Cruciferae). The sterile hybrid resulting from crossing the two genera had 18 chromosomes, nine from the radish and nine from cabbage. The fertile *Raphanobrassica* had 36 chromosomes, including nine pairs of radish chromosomes and nine pairs of cabbage chromosomes.

Speciation in *Tragopogon*

Tragopogon is a genus of perhaps 50 closely related species native to Eurasia and northern Africa. The genus consists of biennial or perennial lactiferous herbs with a well-developed taproot and alternate, linear, entire, grasslike leaves. Plants of the genus produce large, solitary heads at the ends of branches; flowers are perfect and have ligules that are characteristically yellow or purple. Three species, *T. porrifolius* L., *T. pratensis* L., and *T. dubius* Scop., have been introduced to North America and have become widely established over much of the United States.

T. porrifolius, salsify or vegetable oyster, has purple flowers with the heads borne on peduncles that are strongly inflated near the apex. *T. pratensis* has chrome yellow ligules, and the peduncle is slender and not inflated even in fruit. *T. dubius* has pale, lemon-yellow ligules and the head is borne on a strongly inflated peduncle. All three species are characterized by the diploid chromosome number of 12 (Ownbey, 1950). Ownbey noted that the three species are quite distinct, being different in habit, in various aspects of the leaves, in number of florets per head, in relative lengths of involucral bracts and ligules, in colors of the ligules, and in fruit (achene) characters.

Interspecific Hybridization

Tragopogon provides a classical example of interspecific hybridization that is thoroughly documented in the literature. Ownbey (1950) reviewed that literature, which begins with Linnaeus's experimental production in 1759 of what is probably the first interspecific hybrid ever produced for a scientific purpose, that between *T. pratensis* and *T. porrifolius*. Ownbey documented the origin of two new species of *Tragopogon* by means of amphiploidy (i.e., allotetraploidy) in natural populations within recorded history. He described *Tragopogon mirus* (amphiploid *T. dubius* × *T. porrifolius*) and *Tragopogon miscellus* (amphiploid *T. dubius* × *T. pratensis*) and their spontaneous origin in overlapping natural populations of the two respective parental species.

Populations of hybridizing *Tragopogon pratensis* and *T. porrifolius* have been found in Muncie, Indiana. As was indicated above, these species are readily distinguishable, the former being characterized by chrome-yellow ligules and the latter by purple ligules. The interspecific hybrid between *T. pratensis* and *T. porrifolius* is characterized by ligules that have both purple anthocyanin pigment and yellow chromoplasts, giving the ligule of the hybrid a rust or reddish brown color (see Mertens, 1972, for a color photograph). The anthocyanin pigment is restricted to the distal portion of the ligule, whereas the proximal portion of the ligule is yellow. The hybrid head has a yellow center, caused by bicolored ligules. To date, a fertile amphiploid of *T. pratensis* × *T. porrifolius* has not been found in the Muncie populations of plants used for class study.

Pollen Viability

The meiotic process often breaks down in interspecific hybrids because the chromosomes of the two species are not homologous and cannot synapse properly in prophase I of meiosis, leading to very irregular chromosome distribution during the first meiotic division. The consequence of this meiotic disturbance is a high degree of sterility in the interspecific hybrid. Such sterility is ultimately evidenced by the failure of the hybrid to set seed. It can also be detected by the fact that most of the pollen grains produced by the hybrid are defective. These defective pollen grains can be detected by appropriate staining, followed by microscopic examination.

To establish a baseline of fertility, the parent species should also be examined to determine their degree of fertility. Certain environmental conditions (e.g., high temperature, low humidity, drought) may reduce the level of viability in normally fertile plants.

Pollen viability can be determined by staining with cotton-blue in lactophenol. Viable pollen grains stain a uniform dark blue, whereas defective grains stain lightly and appear vacuolated, plasmolyzed, and atypical in size and shape. (*Note:* To make cotton-blue stain, combine 20 ml melted phenol crystals, 20 ml 85% lactic acid, 40 ml glycerine, 20 ml distilled water, and 5 ml 1% aqueous solution of aniline blue.)

Prepare three microscope slides as follows: Place a drop of cotton-blue stain on each microscope slide. On slide 1 dissect a floret of T. porrifolius, freeing pollen grains from the anther. Repeat this procedure for T. pratensis on slide 2, and on slide 3 dissect a floret of the hybrid. Apply a coverslip to each slide. Set the slides aside for 30–60 minutes so that the stain can be absorbed by the pollen grains. At the end of this time, examine each slide at 100× and 430×. [*Note:* Cotton-blue staining for pollen viability may be performed using florets removed from dry herbarium specimens or from flower heads preserved in Carnoy's solution (6:3:1 mixture of absolute ethyl alcohol, chloroform, and glacial acetic acid).]

Count about 100 pollen grains on each slide and record the numbers of viable and nonviable grains in Table S.1. Calculate the percentage viability for the two species and their hybrid. If you had

TABLE S.1. Viability of Pollen Grains Obtained from *T. pratensis, T. porrifolius,* and Their Hybrid as Determined by Cotton-Blue Staining

Species	Total Number of Pollen Grains	Total Number of Viable Grains	Percentage Viability
T. pratensis			
T. porrifolius			
T. pratensis × *porrifolius*			

the amphidiploid derivative of the interspecific hybrid, what level of pollen fertility would you expect in it? Why? How can a "sterile" interspecific hybrid produce an offspring, a fertile amphidiploid? Explain with diagrams.

REFERENCES

MERTENS, T.R. 1972. student investigations of speciation in *Tragopogon. Journal of Heredity* 63(1):39–41+cover.

OWNBEY, M. 1950. Natural hybridization and amphiploidy in the genus *Tragopogon. American Journal of Botany* 37:487–499.

SUPPLEMENT 2

Chromosome Translocations in *Rhoeo spathacea*

The tropical monocot *Rhoeo spathacea* has a low chromosome number ($2n = 12$) and relatively large chromosome size. Each of its 12 chromosomes is involved in a translocation with two other chromosomes in the complement, producing unusual chromosome arrangements in prophase and metaphase of meiosis I and errors in segregation at anaphase I.

Rhoeo is easy to cultivate in the greenhouse (or as a houseplant), and it produces flowers throughout the year, thereby providing a steady supply of material for cytological study. Immature inflorescences (no blooming flowers) should be harvested, preserved in Carnoy's solution, and stored under refrigeration in 70% alcohol. Microsporogenesis in *Rhoeo* can then be studied using the acetocarmine squash procedure outlined in Investigation 6. The meiotic chromosomes of *Rhoeo* resemble those of *Tradescantia* in size and shape and are easy to study and count.

Because of the errors in chromosome segregation produced by the multiple translocations in *Rhoeo*, pollen viability is reduced. For example, one may expect to see 7:5, 8:4, and 9:3 segregations and other atypical segregations in addition to the normal 6:6 arrangement. The level of pollen viability can be determined by staining pollen from mature blossoms using cotton-blue in lactophenol, as described in Supplement 1.

Chromosome Behavior in Rhoeo

The 12 chromosomes of *Rhoeo* are divided by their respective centromeres into two arms each. These chromosomes synapse in meiosis in a precise sequence symbolized as follows:

$$eE — ED — Dc — cC — Cb — bB — Ba — aA — Af — fF — Fd — de$$
$$1 \quad 2 \quad 3 \quad 4 \quad 5 \quad 6 \quad 7 \quad 8 \quad 9 \quad 10 \quad 11 \quad 12$$

Arm E of chromosome 1 is homologous to arm E of chromosome 2, and arm e of chromosome 1 is homologous to arm e of chromosome 12. The consequence of the multiple translocations is that, if crossing over occurs between the respective homologous arms, the 12 chromosomes will be joined in a ring of 12 at late prophase I. Failure of synapsis or crossing over at a single point will break the ring into a single chain of 12 chromosomes. If crossing over does not occur between two different chromosome arms, two separate chains of chromosomes (e.g., 9+3, 6+6, 4+8) can result.

When chromosomes align properly at the equator of the spindle at metaphase I, all the odd-numbered chromosomes are positioned to migrate to one spindle pole and all the even-numbered chromosomes migrate to the opposite pole, as follows:

The even-numbered chromosomes constitute the α complex, and the odd-numbered chromosomes constitute the β complex. All *Rhoeo* plants seem to have resulted from the fusion of a gamete containing the α complex with one containing the β complex. No homozygotes (α/α or β/β) are produced, suggesting that such zygotes are nonviable.

Study of microsporogenesis in *Rhoeo* provides students with the opportunity to observe some of the consequences of chromosome aberrations on the meiotic process. Among the atypical segregations at anaphase I are chromosome bridges, lagging segregation of chromosomes with the formation of micronuclei at telophase I, and atypical segregations, such as 7:5, 5:2:5, or 8:4.

REFERENCES

BAKER, R.F., and T.R. MERTENS. 1975. Meiosis in variegated and anthocyaninless varieties of *Rhoeo*. *Journal of Heredity* 66(6):381–383.
MERTENS, T.R. 1973. Meiotic chromosome behavior in *Rhoeo spathacea*. *Journal of Heredity* 64(6):365–368.
SATTERFIELD, S.K., and T.R. MERTENS. 1972. *Rhoeo spathacea:* a tool for teaching meiosis and mitosis. *Journal of Heredity* 63(6):375–378.

SUPPLEMENT 3

Human Pedigree Analysis

Human pedigrees have been presented for study and analysis in several investigations in this manual. In this supplement, a more generalized approach to pedigree analysis is suggested.

Pedigree analysis is the starting point for all subsequent studies of genetic conditions in families and has been the chief method of genetic study in human kindreds. Students can increase their understanding of the transmission of human traits by studying real pedigrees to be found in most general and all human genetics textbooks. At least eight types of single-gene inheritance can be analyzed in human pedigrees. Listed in Table S.3 are key features and representative examples of these eight mechanisms of inheritance.

Analysis reveals that real pedigrees often cannot be restricted to a single mechanism of inheritance, possibly because the pedigree is too limited in the number of generations or individuals included. In some cases, one may find that none of the eight mechanisms of inheritance fits the pedigree. For example, if the trait under consideration is due to the interaction of two gene loci or is polygenic, then probably none of the single-gene explanations will be able to account for the pattern

TABLE S.3. Key Features and Representative Examples of Eight Single-Gene Mechanisms of Inheritance

Mechanism	Key Features	Examples and McKusick Numbers
Autosomal dominant	Both sexes affected; unaffected parents cannot have affected children; trait does not "skip" generations	Ability to taste PTC (171200) Achondroplasia (100800) Free earlobes (128900)
Autosomal recessive	Both sexes affected; unaffected parents can have affected children; if both parents are affected, all children will be; generations often skipped	Cystic fibrosis (219700) Dry, crumbly earwax (117800) Phenylketonuria (261600)
X-linked dominant	More females affected than males; affected males pass trait to all daughters but to no sons; unaffected parents cannot have affected children	Hypophosphatemia (307800) XG blood group system (314700)
X-linked recessive	More males affected than females; affected females will pass trait to all sons; affected fathers cannot pass trait to sons; affected sons may be produced by normal (carrier) mother and normal father	Hemophilia A (306700) Duchenne muscular dystrophy (310200) Red/green colorblindness (303800 and 303900)
Y-linked	All sons of affected fathers will be affected; females not affected and cannot be carriers	Testis determining factor (480000)
Sex-limited	Traits found in males only or females only	Precocious puberty, male limited (176410)
Sex-influenced male dominant	More males affected than females; all sons of an affected mother will be affected; an affected father can transmit trait to a son; two unaffected parents cannot have an affected daughter	Male pattern baldness (109200) Short index finger (136100)
Sex-influenced female dominant	More females affected than males; all daughters of an affected father will be affected; two unaffected parents cannot have an affected son	Heberden nodes (140600)

of inheritance in the pedigree. Furthermore, if the trait has no genetic basis at all, then none of the eight mechanisms would be expected to explain the pedigree pattern.

The references by Nussbaum et al. (2004) and Mertens (1990) will be useful in getting started in analyzing pedigrees. Look for critical matings that exclude the mechanism of inheritance. For example, if a woman expresses a certain abnormality and her sons do not, one can reasonably exclude X-linked recessive inheritance. Second, think in concrete terms when analyzing pedigrees. For example, if you are trying to decide whether or not the phenotype in question is an autosomal dominant trait, ask yourself the question, "Could this be a pedigree of Huntington's disease (a known autosomal dominant)?" Next, look at the pedigree with a specific autosomal recessive trait (e.g., cystic fibrosis) in mind. Continue with other specific examples of the various types of inheritance.

Furthermore, to facilitate pedigree analysis in the classroom, certain assumptions should be made.

1. The phenotype is under exclusive genetic control.

2. There is complete penetrance of the trait under question.

3. Each phenotype has its own unique genetic cause (i.e., there are no "duplicate phenotypes," some caused by autosomal dominants, others by autosomal recessives, etc.).

Although these assumptions would not be warranted in the real world of medical genetics, they would seem to be appropriate in a classroom setting. Keep in mind, however, that in actual practice pedigree analysis can be much more complex than what is experienced in the classroom setting.

REFERENCES

McKusick, V.A. 1998. *Mendelian inheritance in man: a catalog of human genes and genetic disorders*, 12th ed. Baltimore: Johns Hopkins University Press.

Mertens, T.R. 1990. Using human pedigrees to teach Mendelian genetics. *The American Biology Teacher* 52(5):288–290.

Nussbaum, R.L., R.R. McInnes, and H.F. Willard. 2004. *Thompson & Thompson genetics in medicine*, 6th ed. Philadelphia: W. B. Saunders.

Online Mendelian inheritance in Man, OMIM™. McKusick-Nathans Institute for Genetic Medicine, Johns Hopkins University (Baltimore, MD) and National Center for Biotechnology Information, National Library of Medicine (Bethesda, MD), 2000. http://www.ncbi.nih.gov/omim/

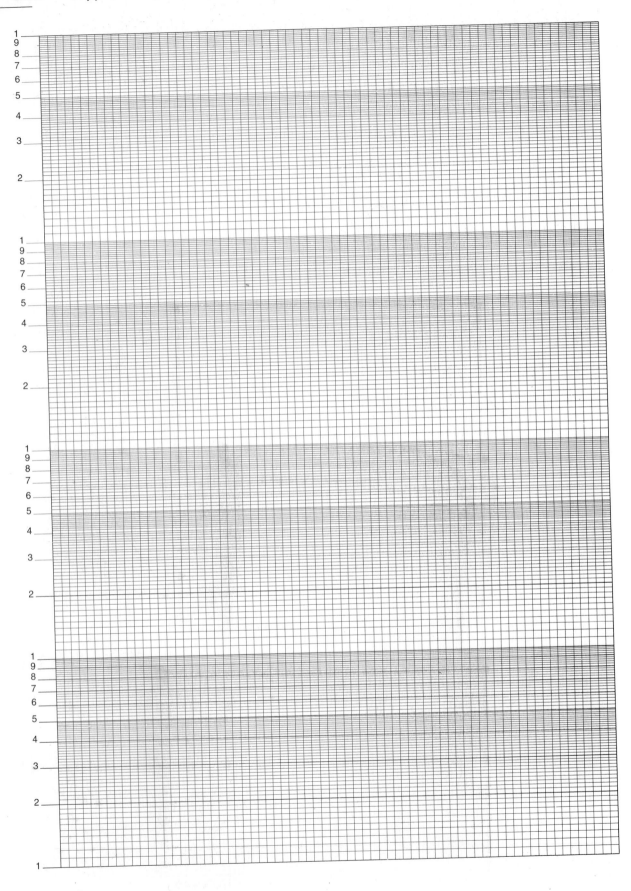